# FBI
## GUIDE TO
# CONCEALABLE WEAPONS

**Paladin Press • Boulder, Colorado**

*FBI Guide to Concealable Weapons*
Copyright © 2004 by Paladin Press

ISBN 1-58160-447-5
Printed in the United States of America

Published by Paladin Press, a division of
Paladin Enterprises, Inc.
Gunbarrel Tech Center
7077 Winchester Circle
Boulder, Colorado 80301 USA
+1.303.443.7250

Direct inquiries and/or orders to the above address.

PALADIN, PALADIN PRESS, and the "horse head" design
are trademarks belonging to Paladin Enterprises and
registered in United States Patent and Trademark Office.

All rights reserved. Except for use in a review, no
portion of this book may be reproduced in any form
without the express written permission of the publisher.

Neither the author nor the publisher assumes
any responsibility for the use or misuse of
information contained in this book.

Visit our Web site at www.paladin-press.com

## FOREWORD

In the wake of the September 11, 2001, airline hijackings, the Firearms and Toolmarks Unit of the FBI Laboratory started a collection of small and easily concealed knives.

This is the first installment of a continuing effort to collect and distribute information on knives that otherwise may be dismissed as nonthreatening items. Many of the knives in this collection were commercially purchased and typically can be bought for less than $20. Some of these knives are common items found in most homes and offices.

You will notice also that some are made of a plastic material, making them less likely to be considered a weapon. Each of these tools was designed to cut and is fully functional in that respect. Whether used to cut paper, cardboard, or other material, these knives should be treated as potentially dangerous weapons. Each knife is shown with an accompanying scale for size reference, and many include an X-ray photograph to show how these weapons might appear if placed in luggage and passed through a scanning device.

Any information concerning these knives, or cutting tools not shown in this booklet, can be forwarded to:

Federal Bureau of Investigation
Laboratory Division
Firearms/Toolmarks Unit
2501 Investigation Parkway
Quantico, VA 22135
(703) 632-7225/6

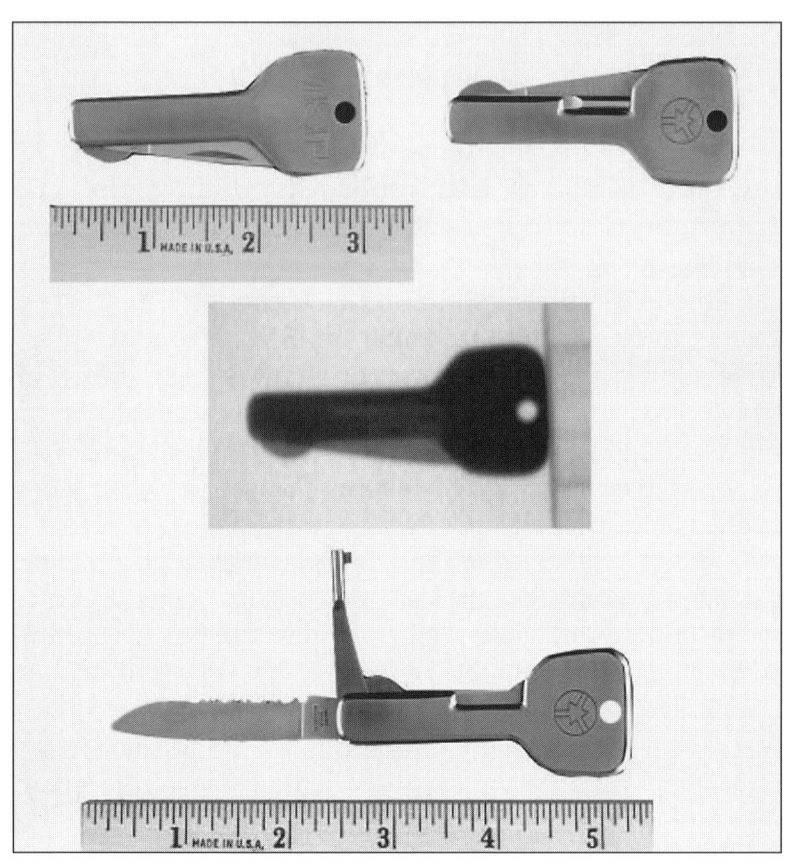

## KEY KNIFE

Manufacturer: ASP
Made in Japan
Composition: Metal
Note: Looks like an oversized key. Stored inside are a blade and a handcuff key.

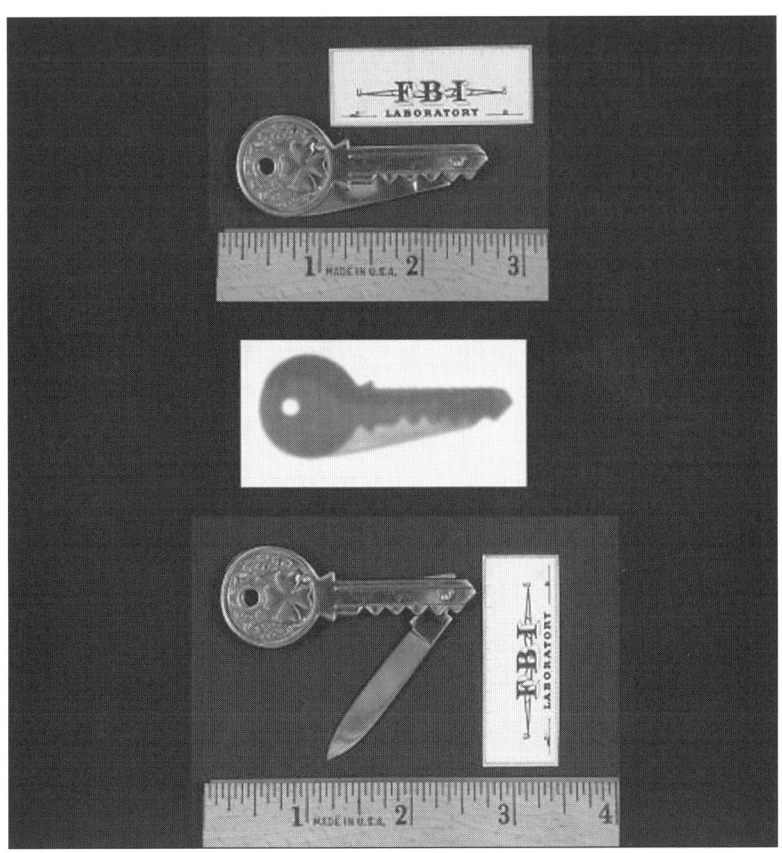

**KEY**

Manufacturer: United Cutlery
Made in Italy
Composition: Metal
Note: Appears to be a key when placed on a ring with others. Blade folds out from end of key.

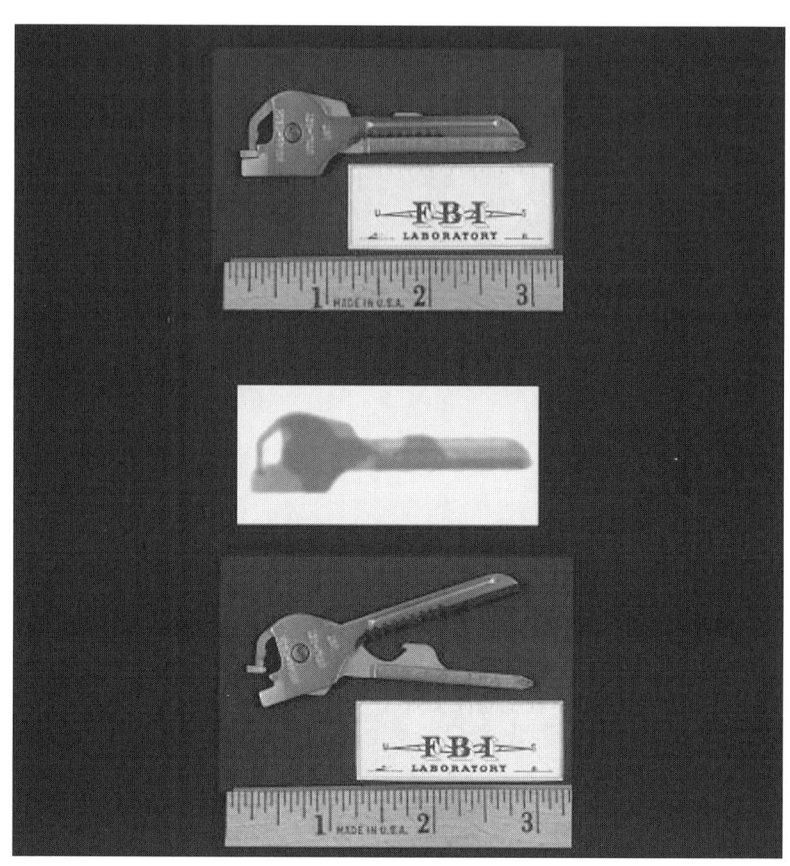

## KEY

Manufacturer: Swiss-Tech
Made in USA
Composition: Metal
Note: Appears to be a key when placed on a ring with others. Opens in a scissorlike fashion to expose knife blade, serrated blade, and flat- and Phillips-head screwdrivers.

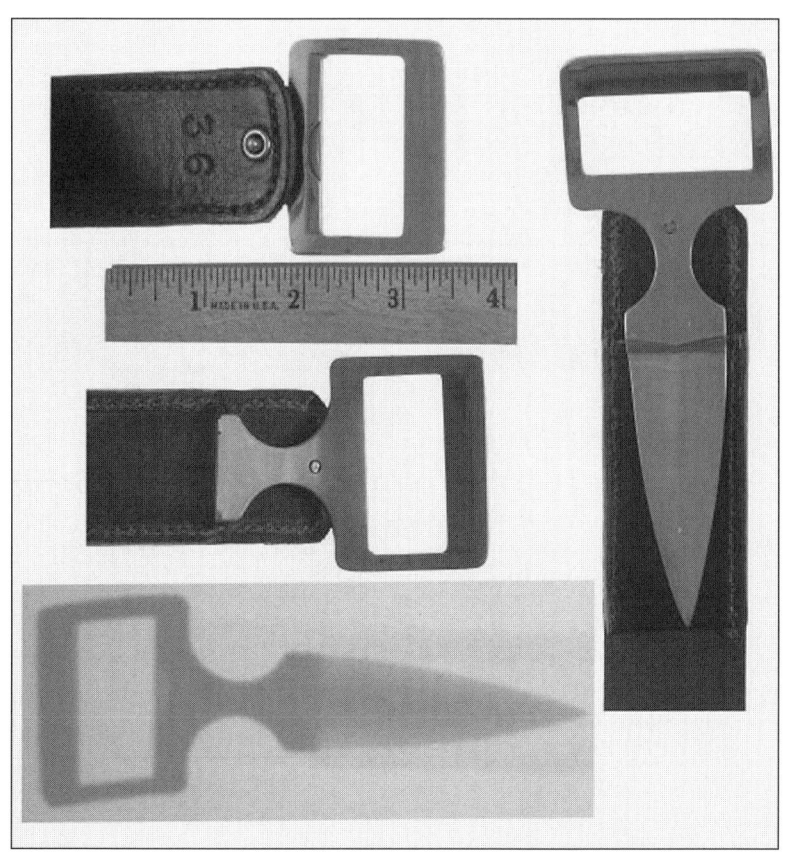

## BOWEN BELT KNIFE

Manufacturer: Bowen Knives
Made in USA
Composition: Metal blade with leather belt
Note: Blade is hidden inside of belt; only the buckle "handle" is seen.

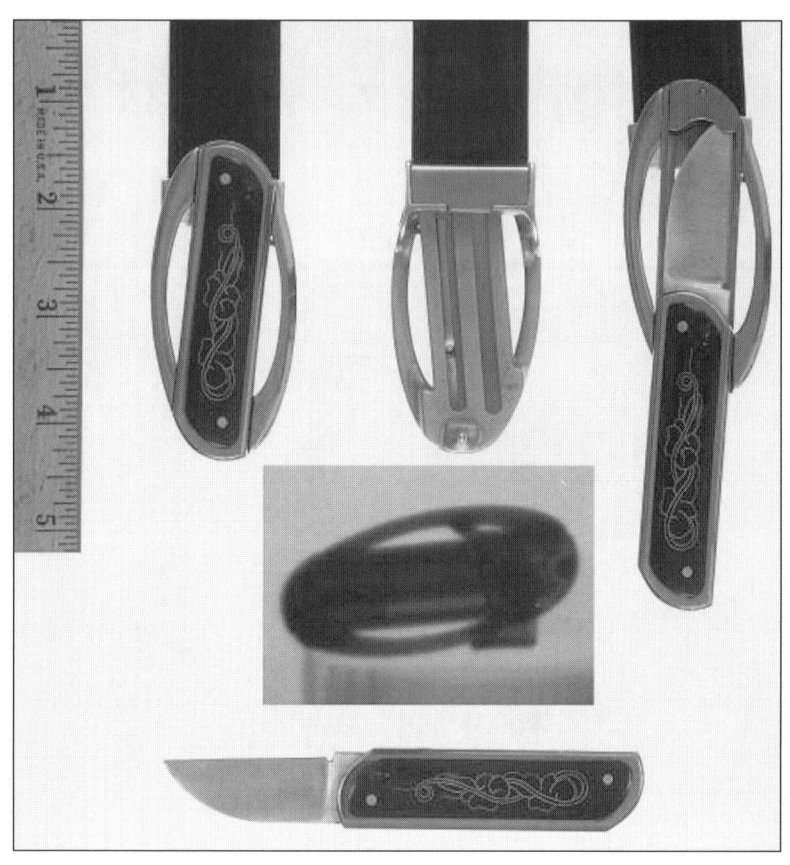

**BELT KNIFE**

Manufacturer: C.A.S. Iberia, Inc.
Made in USA
Composition: Metal and leather
Note: Looks like an ordinary belt buckle. Buckle stores a knife, which can be pulled out to expose the blade.

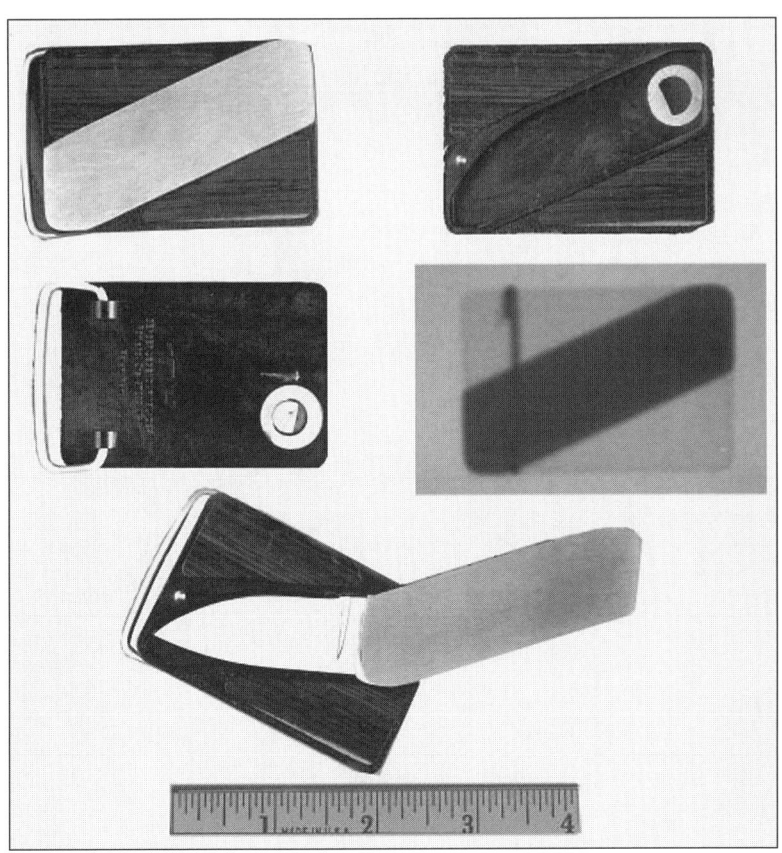

## TOUCHÉ BELT BUCKLE

Manufacturer: Gerber
Made in USA
Composition: Metal
Note: The knife blade folds under the handle, which can be stored in the front of the belt buckle.

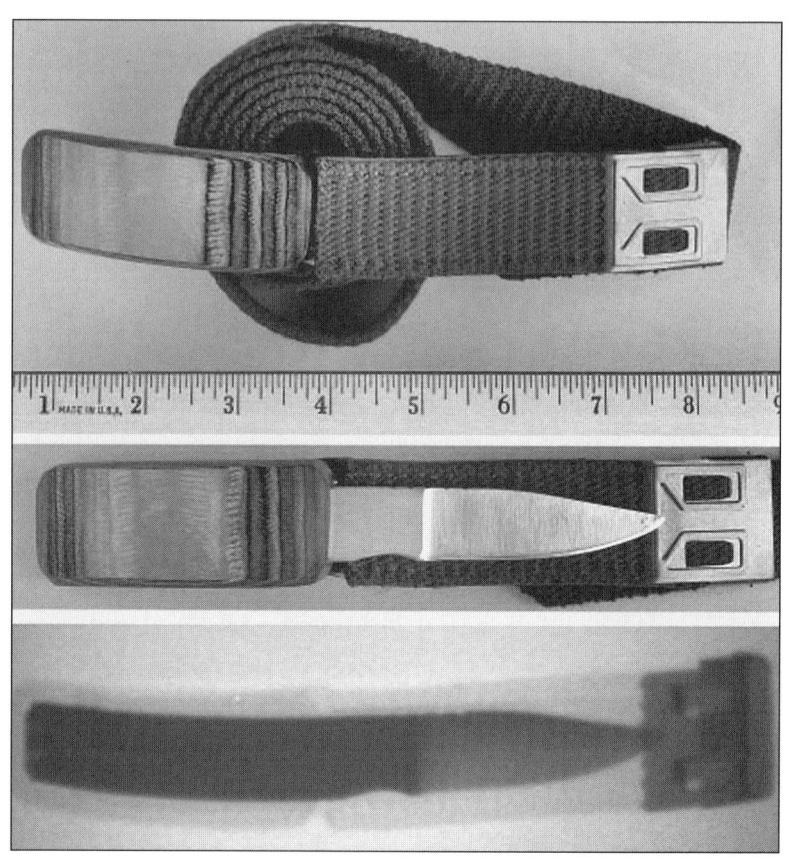

## BELT KNIFE

Manufacturer: Valois Knives
Made in USA
Composition: Metal
Note: Appears as a normal belt. The "buckle" pulls from the belt to expose the blade. Comes in various styles.

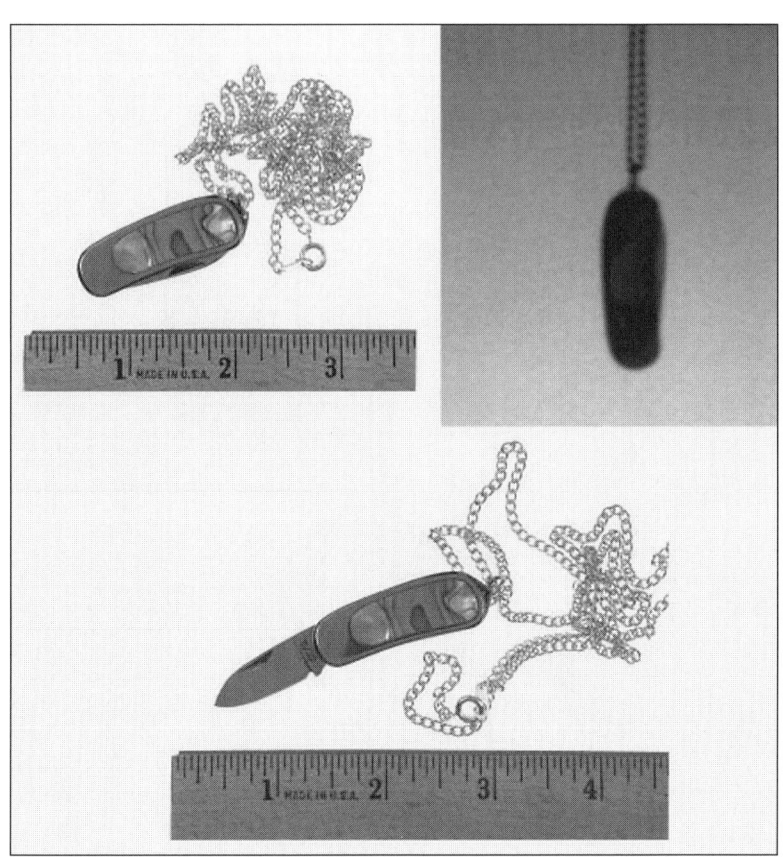

## NECKLACE KNIFE

Manufacturer: Kaicut Knives (model name—Moki)
Made in Germany
Composition: Stainless steel with a decorative inlay
Note: May be worn as a necklace or carried as a
   key chain.

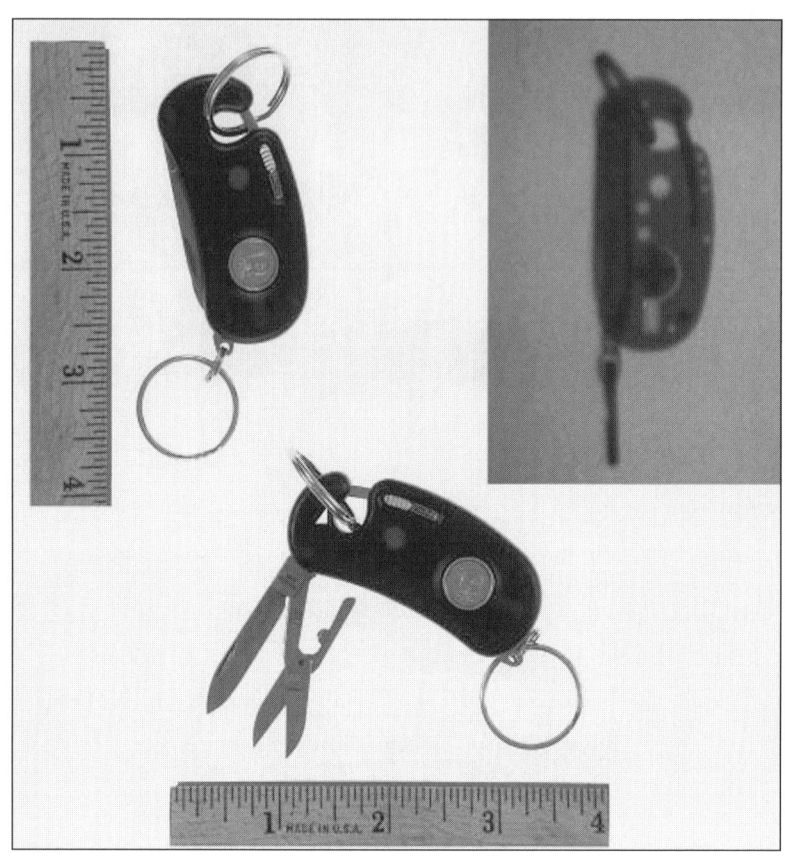

### KEY RING KNIFE

Manufacturer: Marex
Made in Korea
Composition: Plastic and metal
Note: Looks like a key ring but houses a blade, scissors, and flashlight.

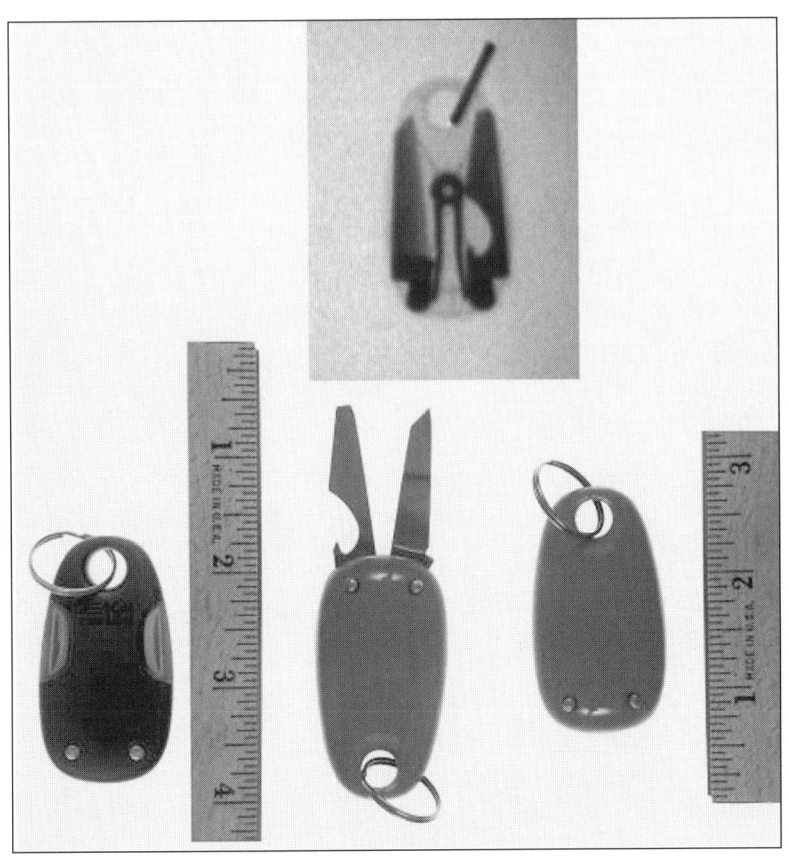

## KEY RING KNIFE

Manufacturer: EKA
Made in Sweden
Composition: Plastic and metal
Note: Looks like a key ring but houses a blade and
   bottle opener.

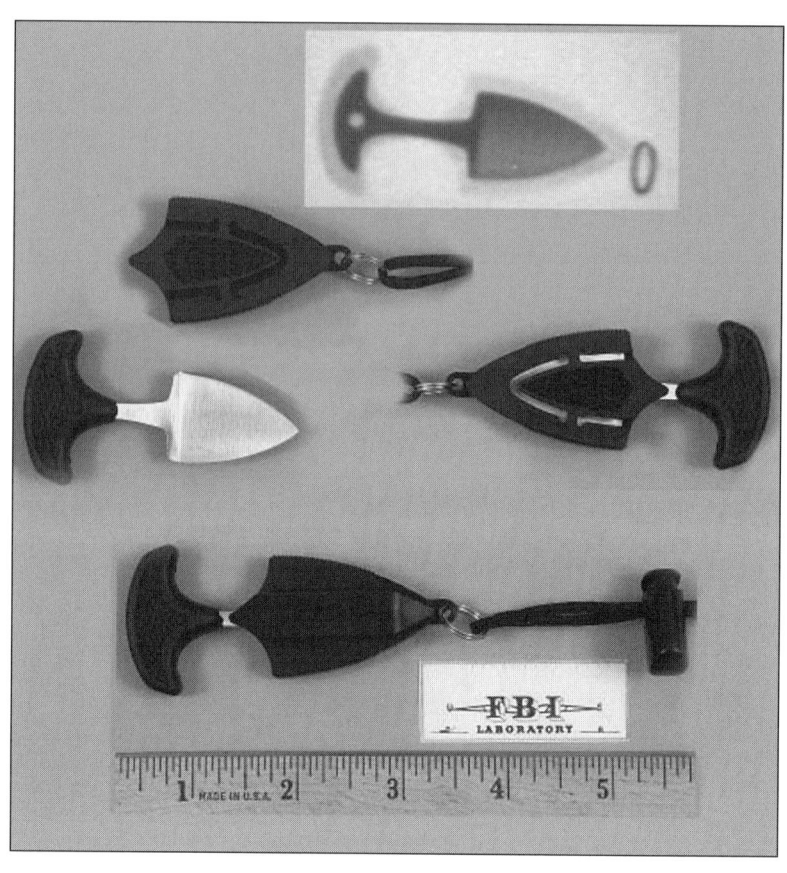

## PUSH DAGGER

Manufacturer: United Cutlery
Made in Taiwan
Composition: Plastic sheath and handle with metal blade
Note: Sheath has a clip allowing it to be worn on belt, boot, or pocket. Also comes with lanyard to wear around neck or possibly use as a key chain.

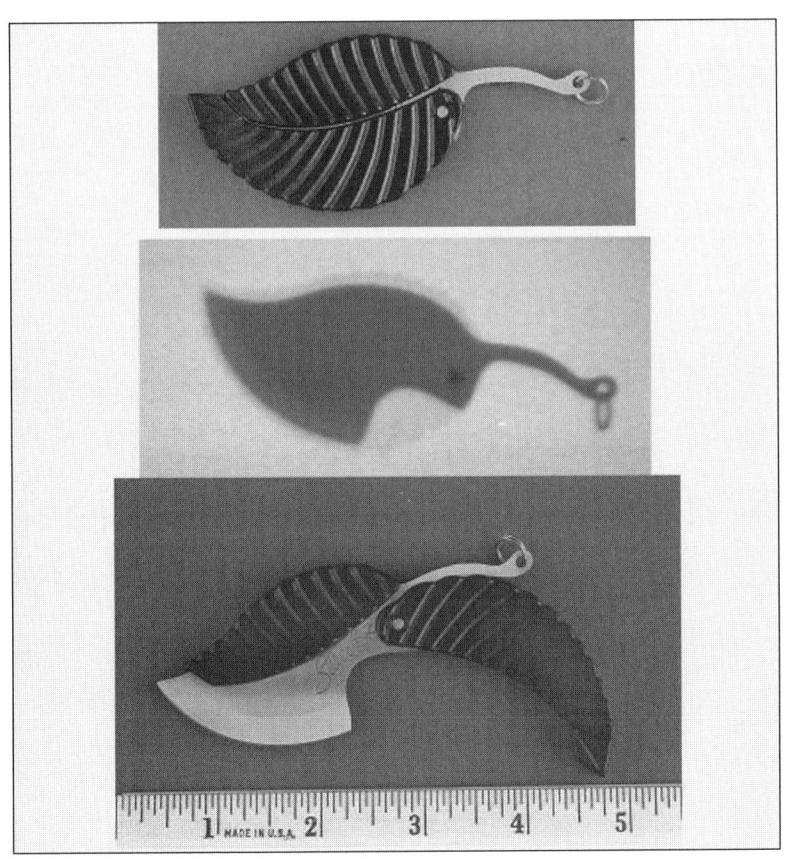

## LEAF KNIFE

Manufacturer: Unknown
Made in USA
Composition: Plastic and metal
Note: Plastic "leaf" opens to expose knife blade. Also comes in an all-stainless model. Can be worn on a neck chain or as a key ring.

## KUBATON

Manufacturer: United Cutlery
Made in Taiwan
Composition: Metal
Note: Styles vary by manufacturer.

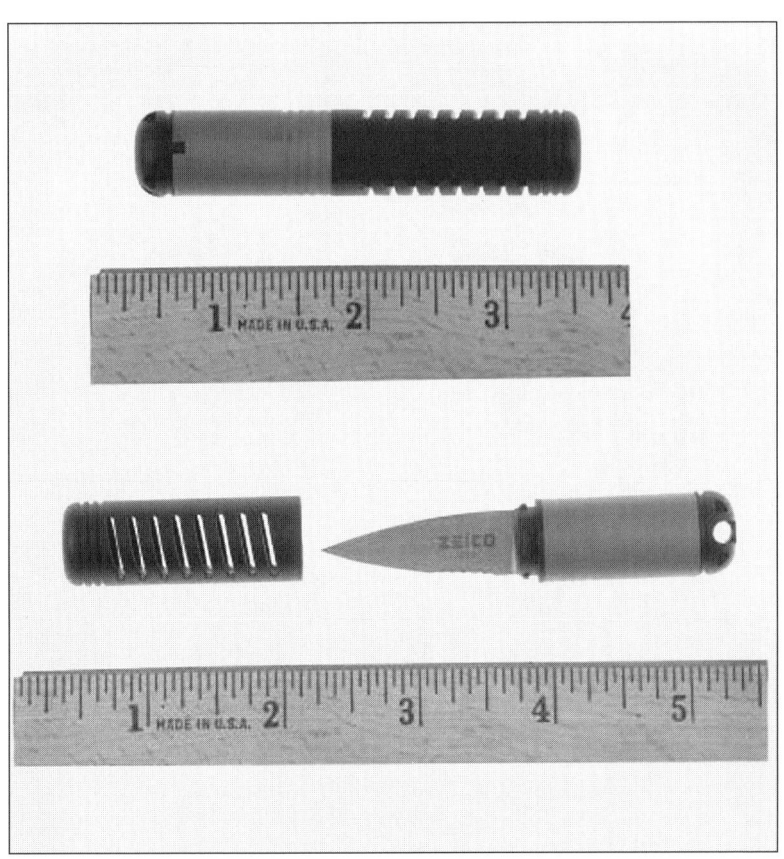

## KEY RING KNIFE

Manufacturer: Zelco
Made in USA
Composition: Metal blade with two-piece plastic handle
Note: Can be used as a key ring or worn on a neck chain.

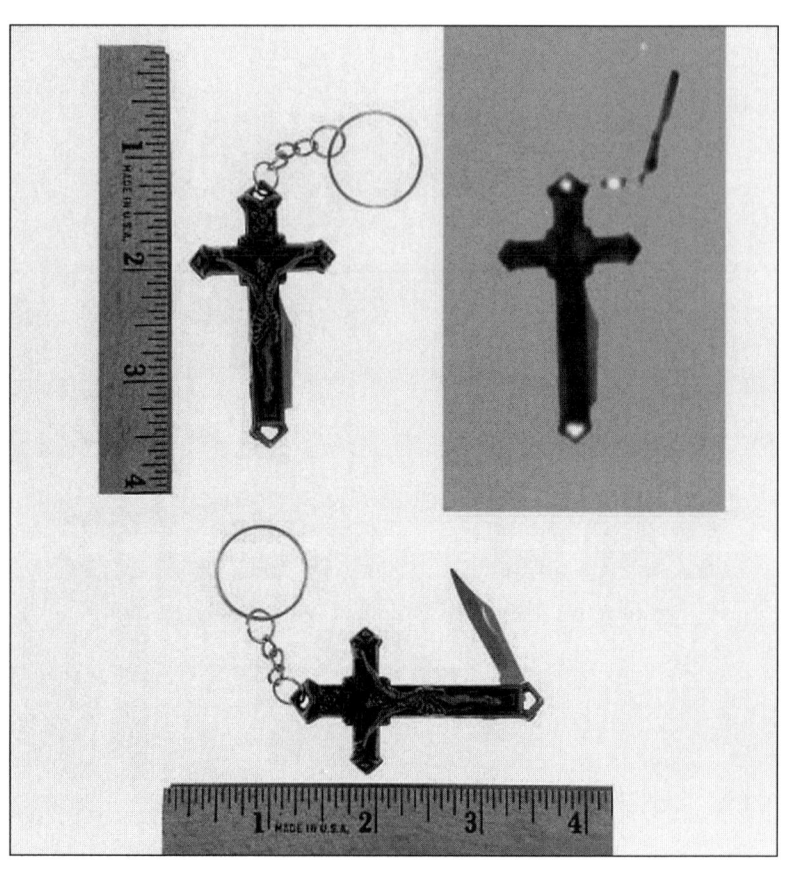

## CRUCIFIX KNIFE

Manufacturer: Unknown
Composition: Nonmagnetic cross with metal blade and key chain
Note: Can be used as a key chain or necklace.

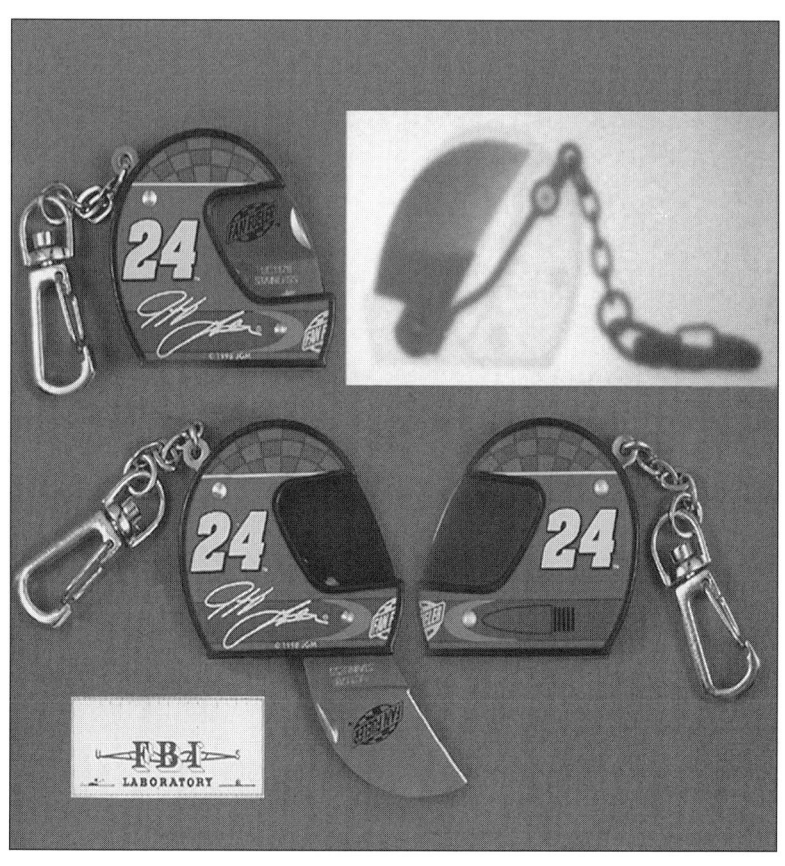

## NASCAR HELMET KNIFE

Manufacturer: Action Racing Collectibles
Made in China
Composition: Plastic case with metal knife blade
Note: Colors and styles vary depending on the NASCAR theme.

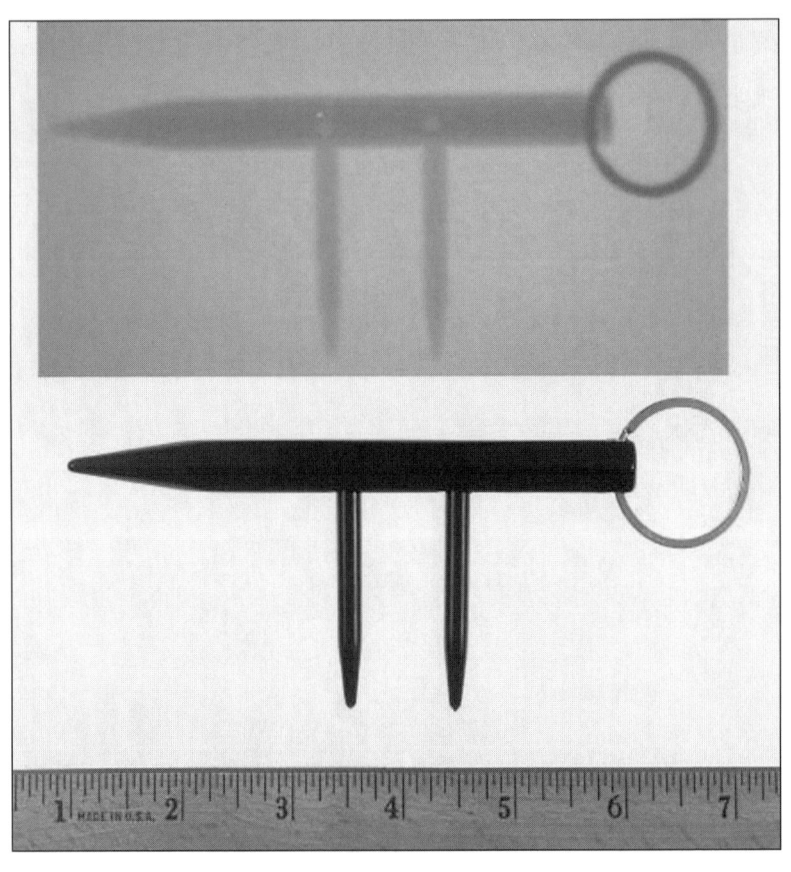

## NINJA KEY RING

Manufacturer: Unknown
Made in USA
Composition: Metal
Note: Impact tool.

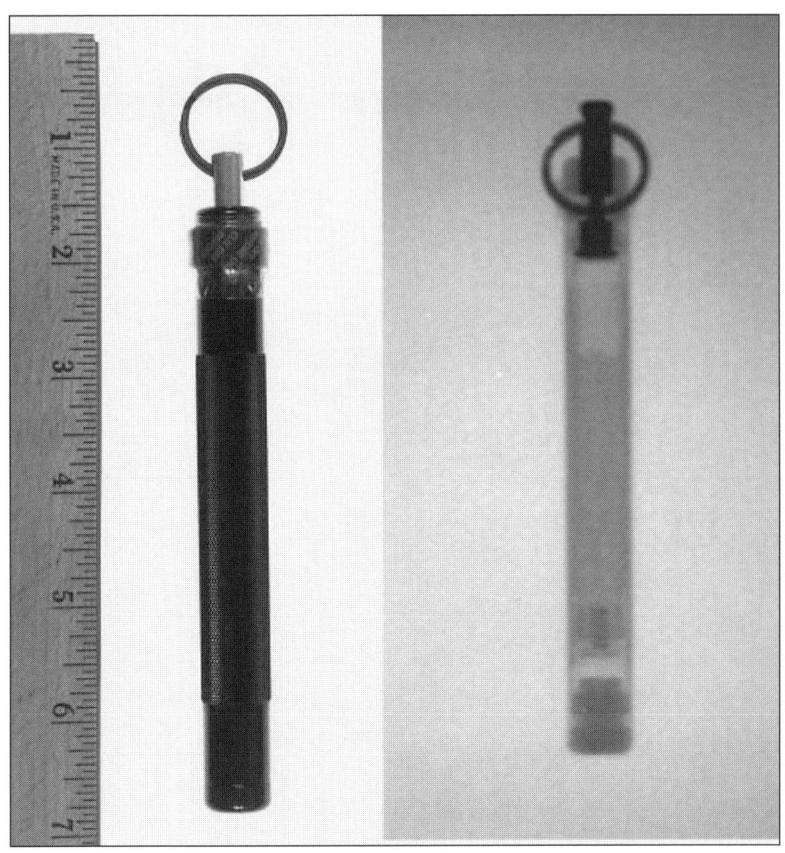

## PEPPER SPRAY KEY RING

Manufacturer: Unknown
Made in Brazil
Composition: Tubular metal body with metal ring
Note: Ejects red pepper spray.

## COIN

Manufacturer: United Cutlery
Made in Taiwan
Composition: Metal
Note: Appears to be a normal silver dollar key chain. This knife may vary in style and blade type.

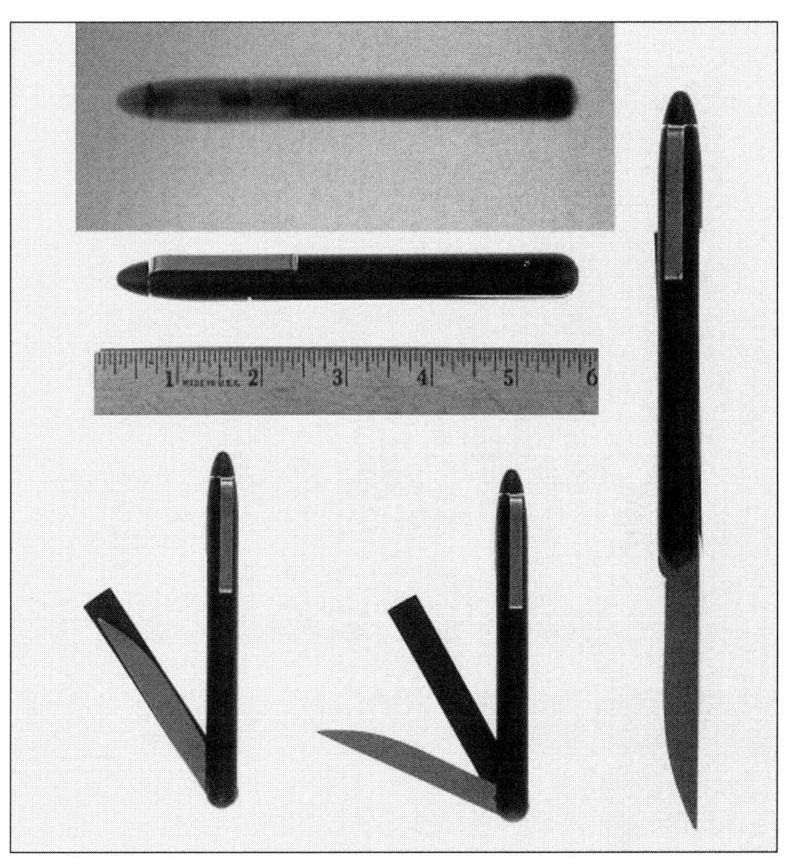

## DELTA PEN

Manufacturer: Unknown
Made in ?
Composition: Metal
Note: Looks like an ink pen and can be opened to expose a knife blade.

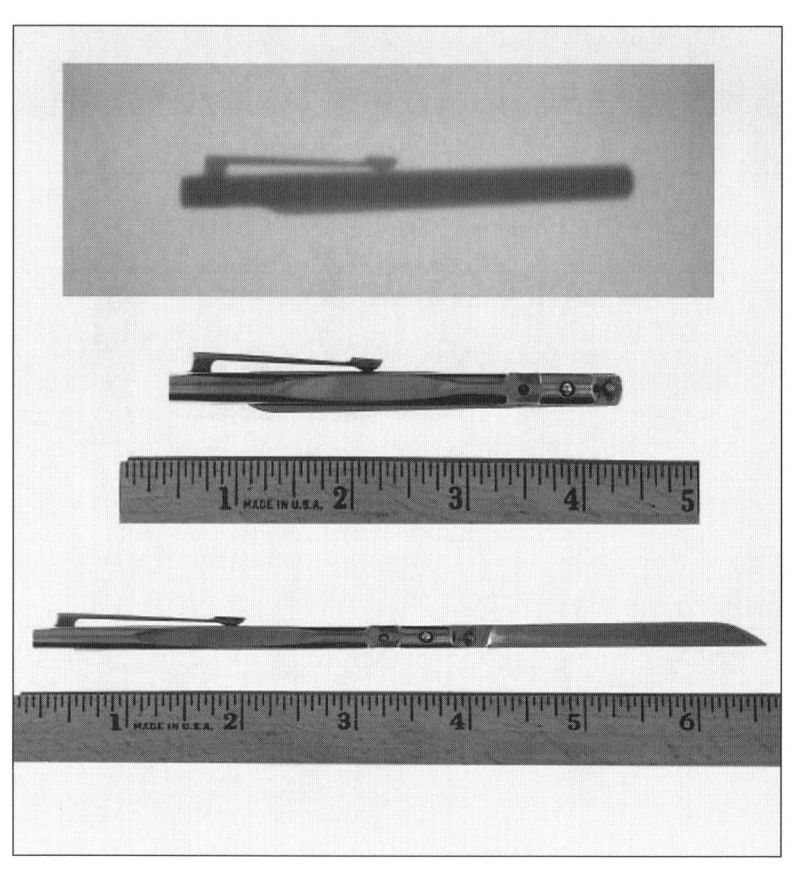

## EXECUTIVE PEN

Manufacturer: Unknown
Made in USA
Composition: Metal
Note: Looks like a pen and can be opened to expose a knife blade. Comes in various sizes.

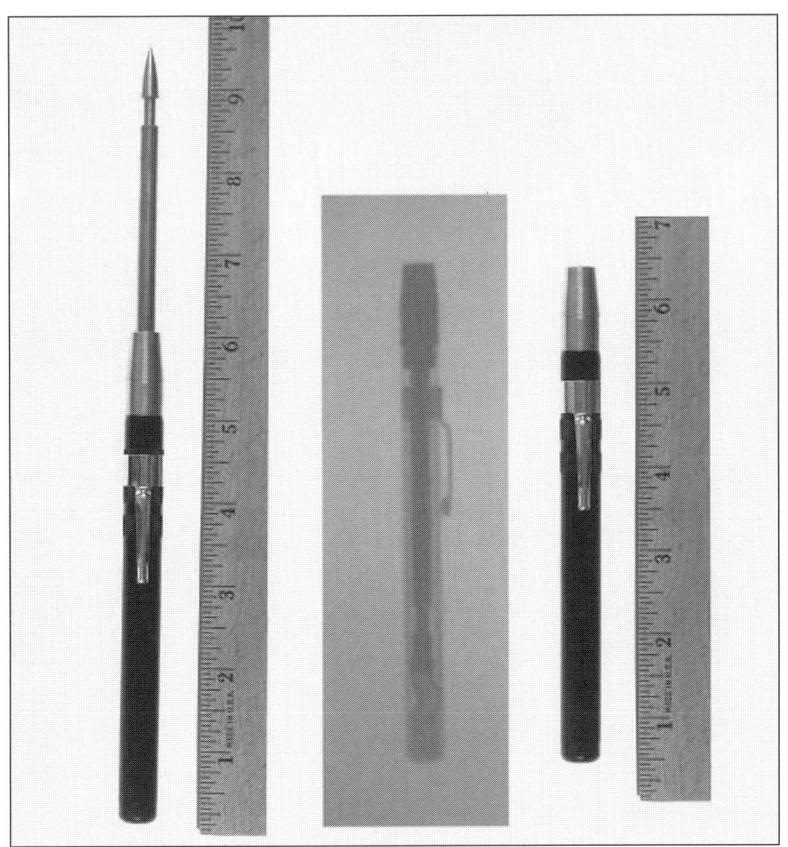

## AUTO SPIKE

Manufacturer: S.W.A.T.
Made in USA
Composition: Metal spike with metal handle
Note: Spring-loaded spike deployed by auto release.

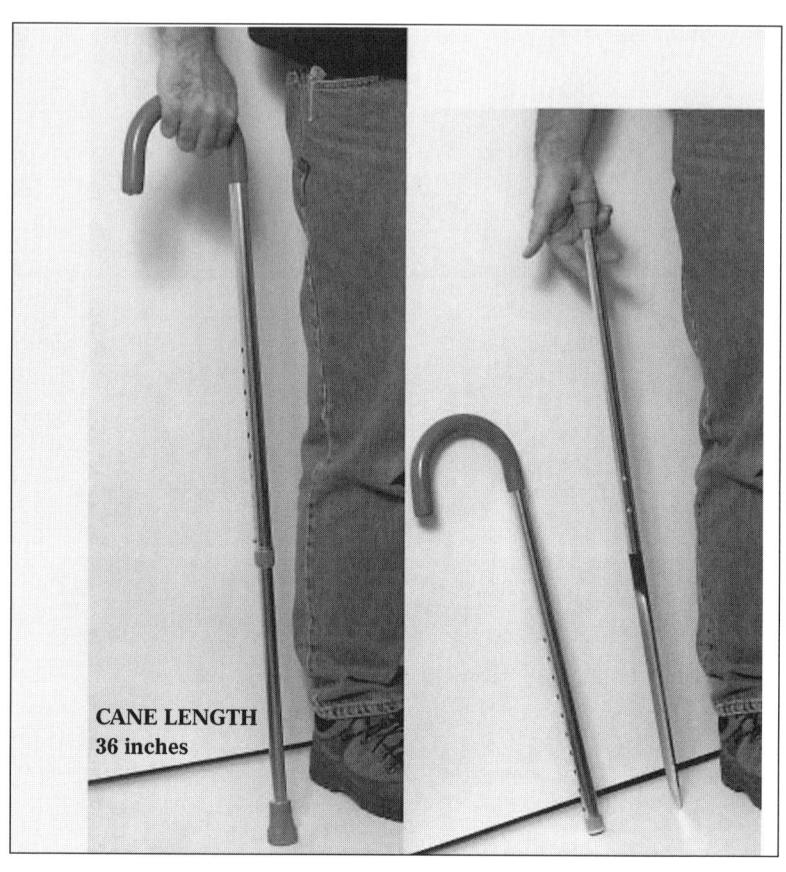

## WALKER SWORD

Manufacturer: Unknown
Made in USA
Composition: Metal
Note: Looks like an ordinary walking cane. Expandable shaft can be removed from handle portion to expose knife blade.

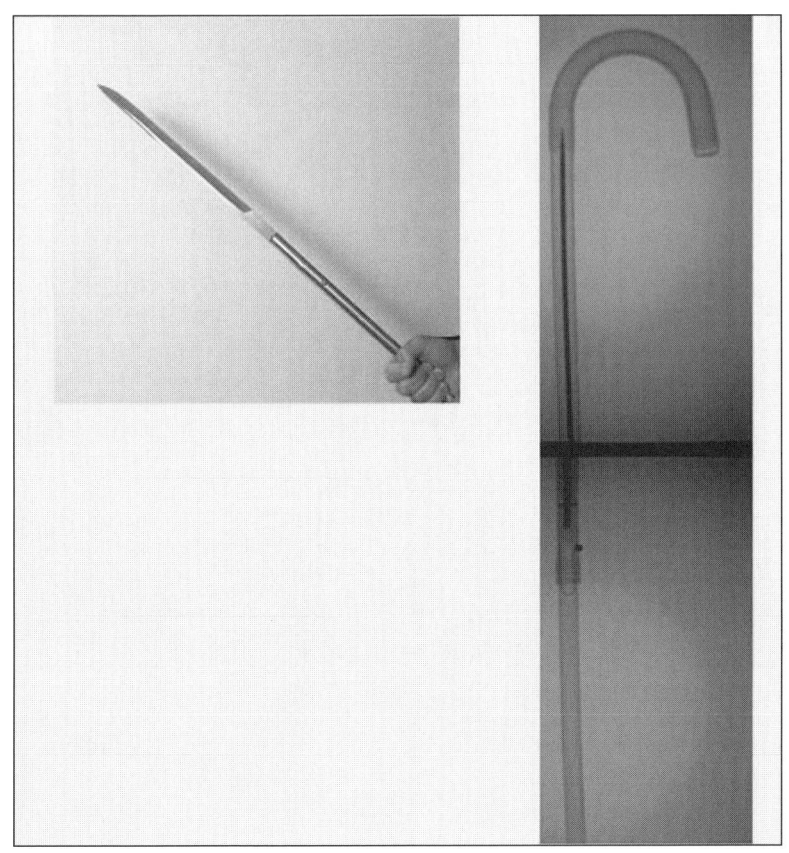

The shaft of this cane is removed to expose the knife blade. The blade could also be fitted in a walker/crutch.

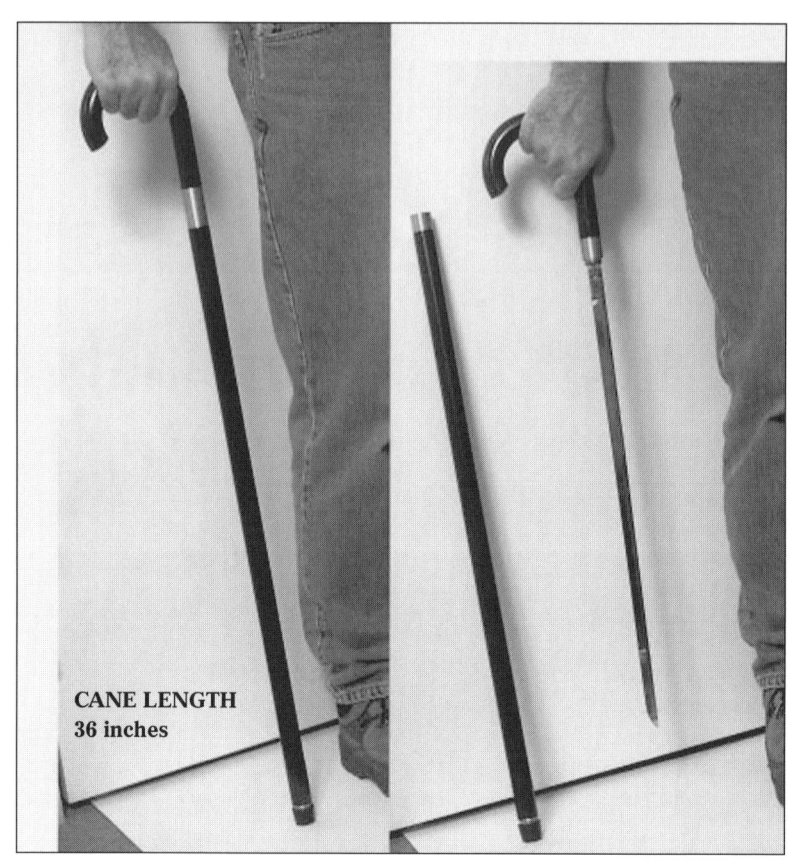

## CANE SWORD

Manufacturer: Unknown
Made in USA
Composition: Metal
Note: Looks like an ordinary cane. Handle unscrews to expose knife blade.

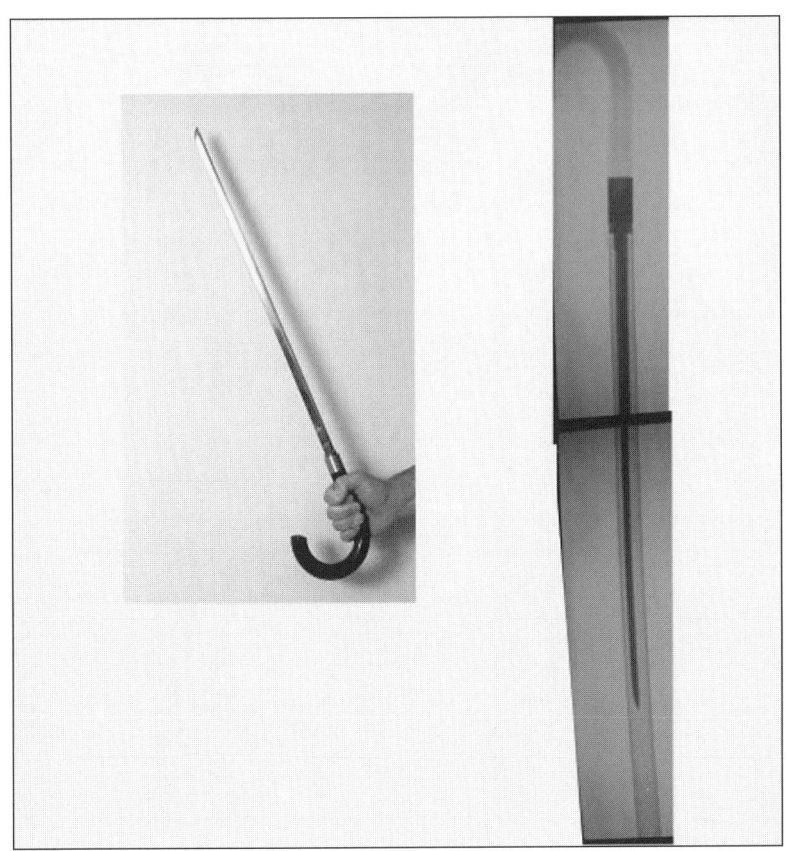

The handle of the cane is removed to expose the knife blade.

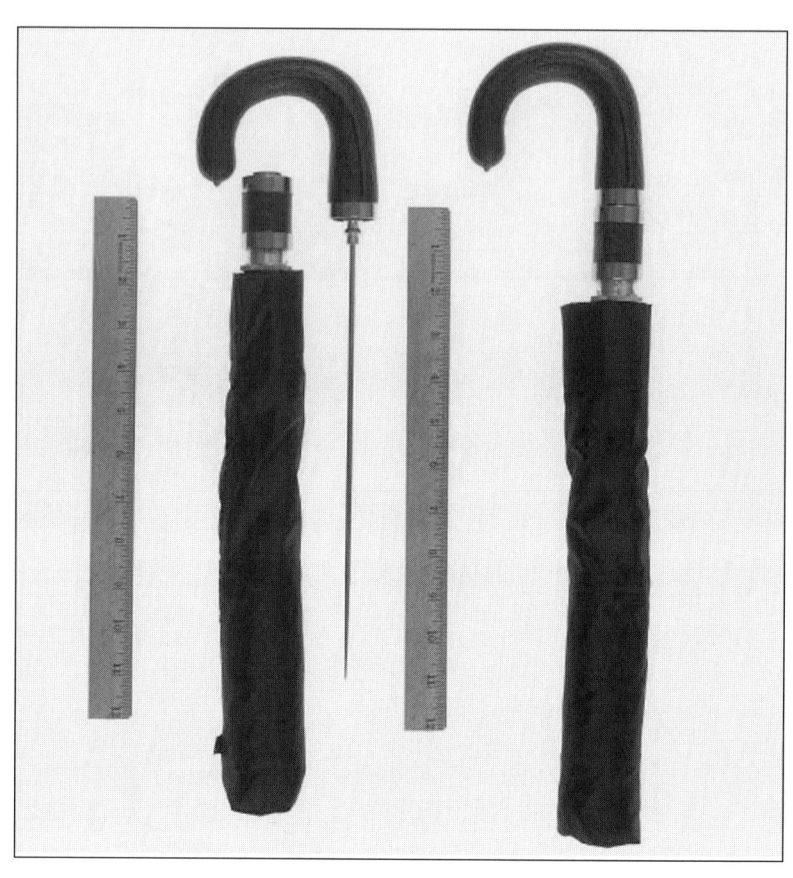

## SWORD UMBRELLA

Manufacturer: Atlanta Cutlery
Made in USA
Composition: Metal and plastic
Note: Appears as a normal folding umbrella. Handle pulls free to expose a 10-inch steel spike.

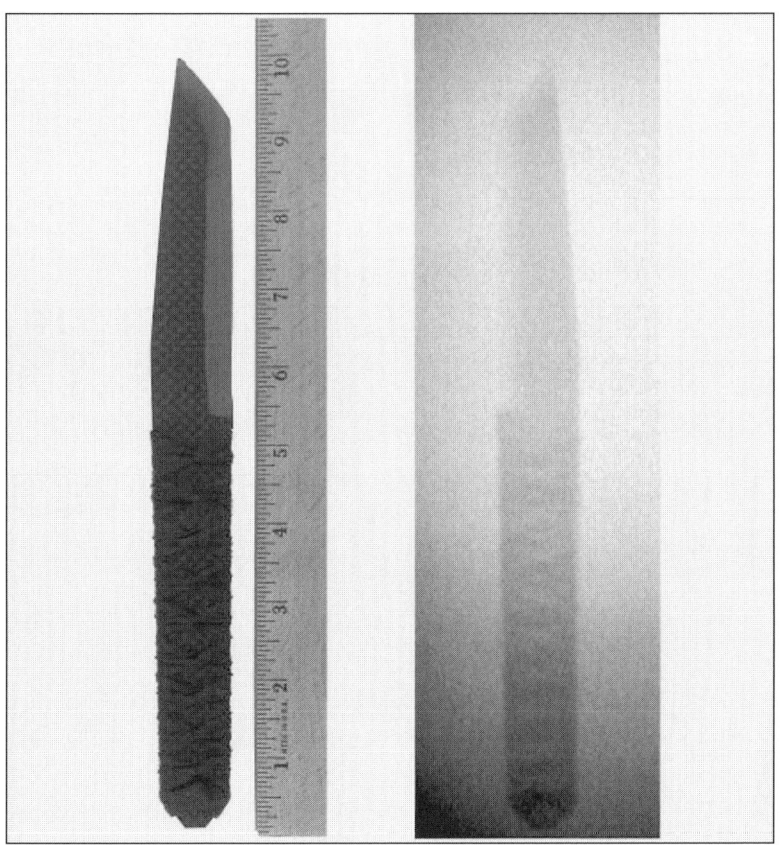

**CARBON FIBER KNIFE**

Manufacturer: Unknown
Made in USA
Composition: Plastic
Note: All-plastic construction makes this invisible to a magnetometer.

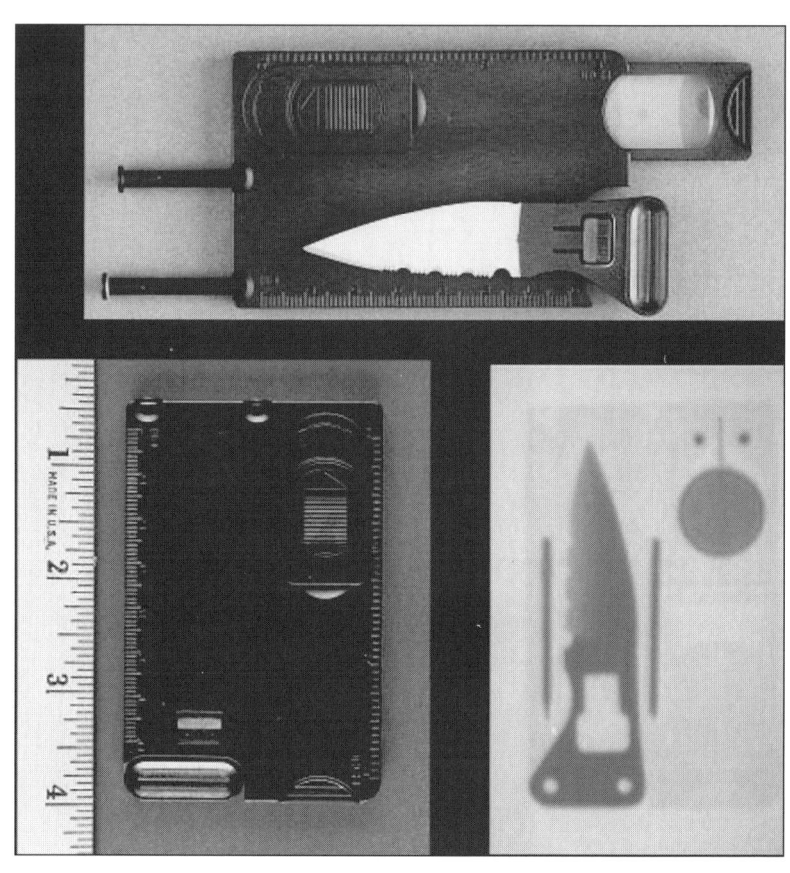

## TOOL CARD

Manufacturer: Unknown
Composition: Plastic with metal knife blade
Note: Designed to be the approximate size of a credit card so it can be carried in a pocket. Includes a magnifying glass, two screwdrivers, and a flashlight.

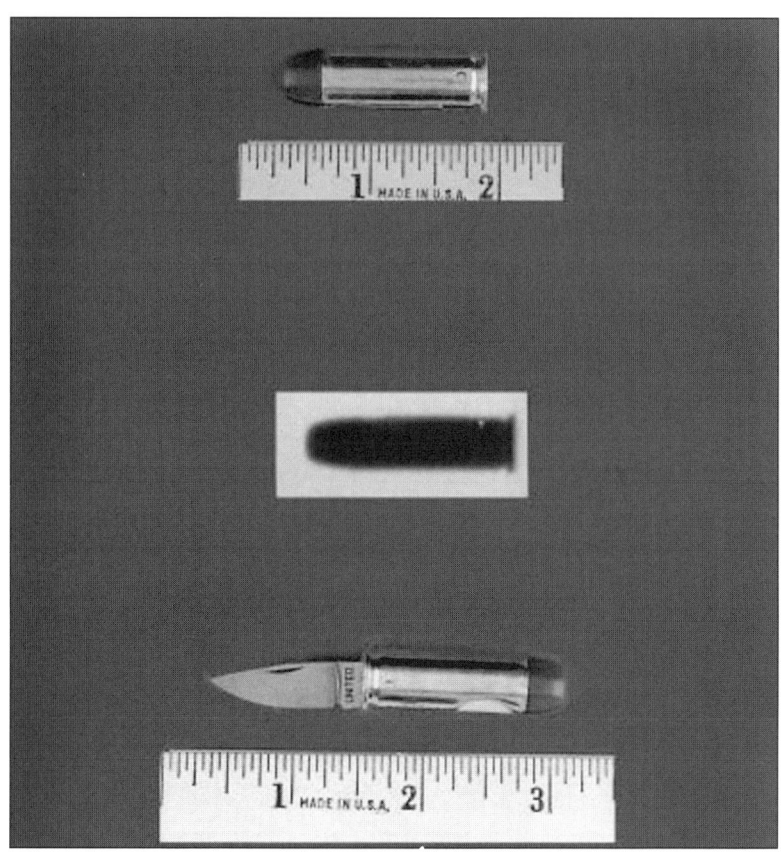

## BULLET (CARTRIDGE) KNIFE

Manufacturer: United Cutlery
Made in Italy
Composition: Metal blade inside of cartridge
Note: Looks like a cartridge for a handgun. Headstamp reads "30-06 SPRG."

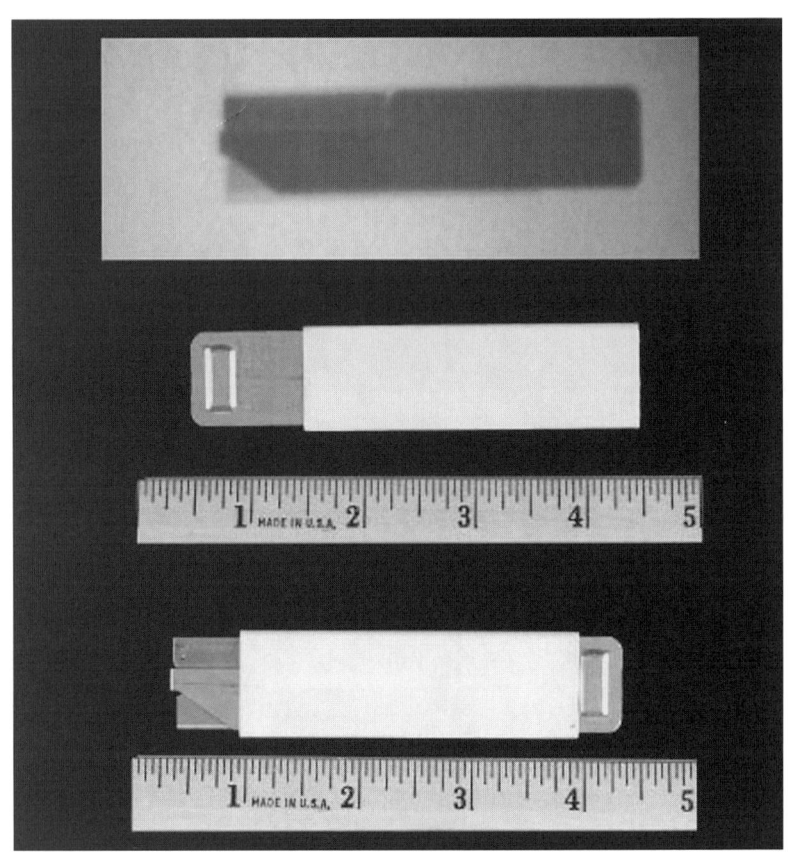

**BOX CUTTER**

Manufacturer: Unknown
Made in USA
Composition: Metal razor blade with metal handle

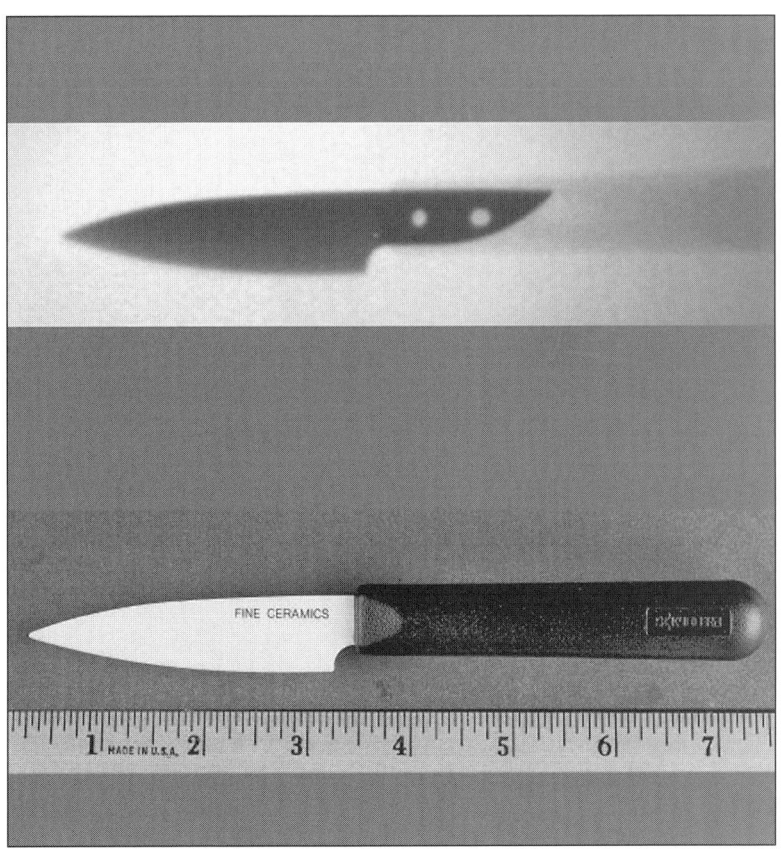

## CERAMIC KNIFE

Manufacturer: Kyocera
Made in Japan
Composition: Ceramic blade with plastic handle
Note: Ceramic and plastic construction makes it invisible to a magnetometer.

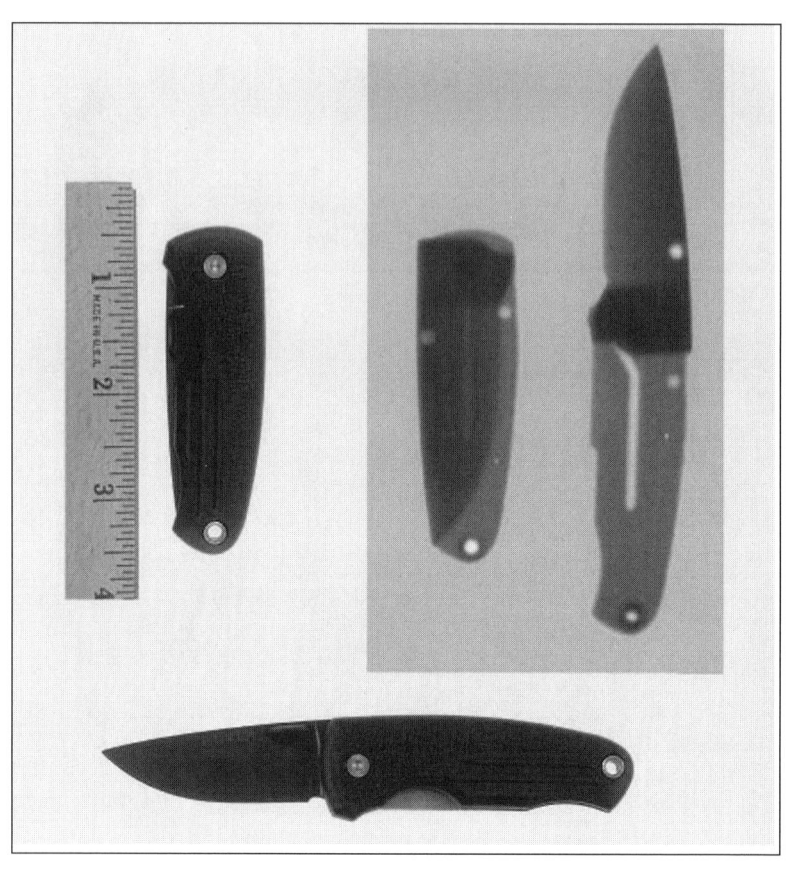

**CERAMIC FOLDING KNIFE**

Manufacturer: Boker Knives (model name—Gamma)
Made in Germany
Composition: Plastic case with ceramic blade and metal rivets and liner lock
Note: Minimal amount of metal makes this knife invisible to a magnetometer.

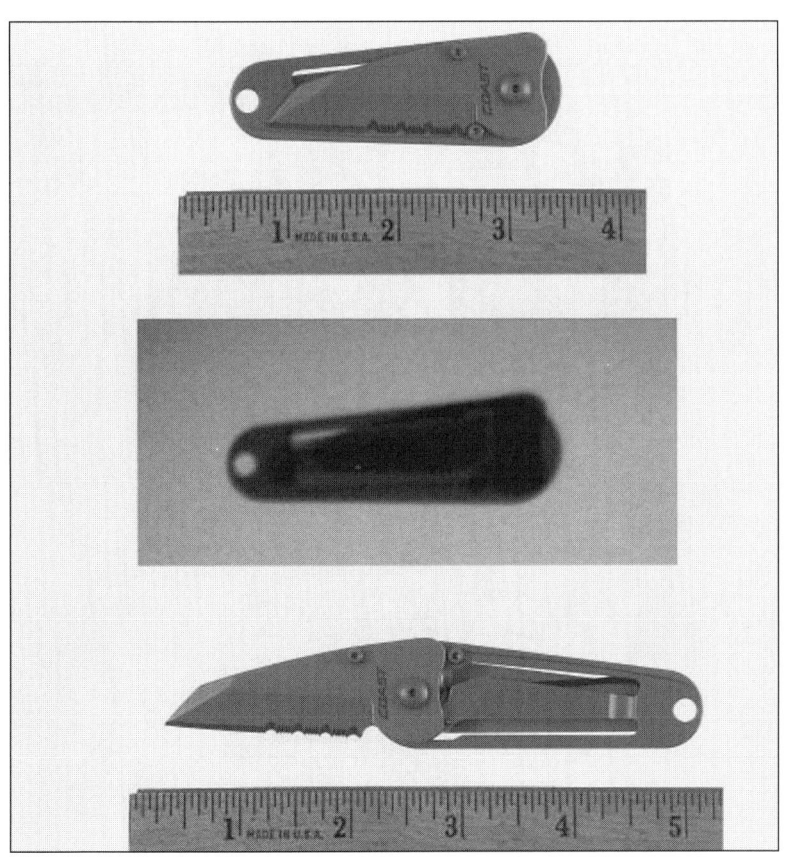

## FOLDING KNIFE

Manufacturer: Coast Knives
Made in Taiwan
Composition: Metal blade and handle
Note: Small folding knife with pocket/money clip.

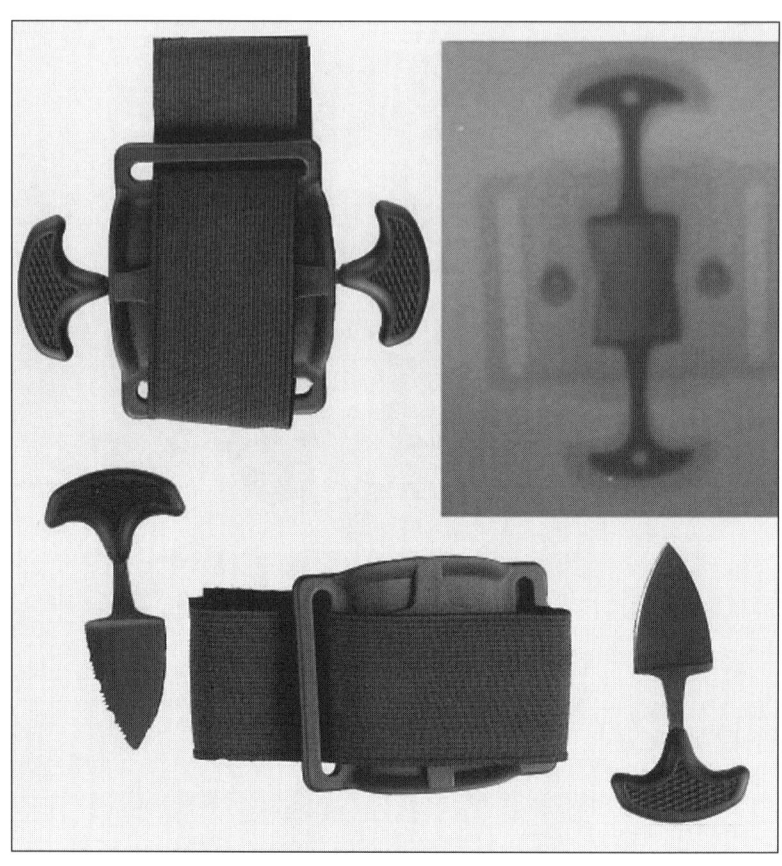

## DOUBLE-HANDED DAGGER

Manufacturer: Unknown
Made in ?
Composition: Metal blade with plastic handle
Note: Can be worn on belt or with wrist/ankle strap.

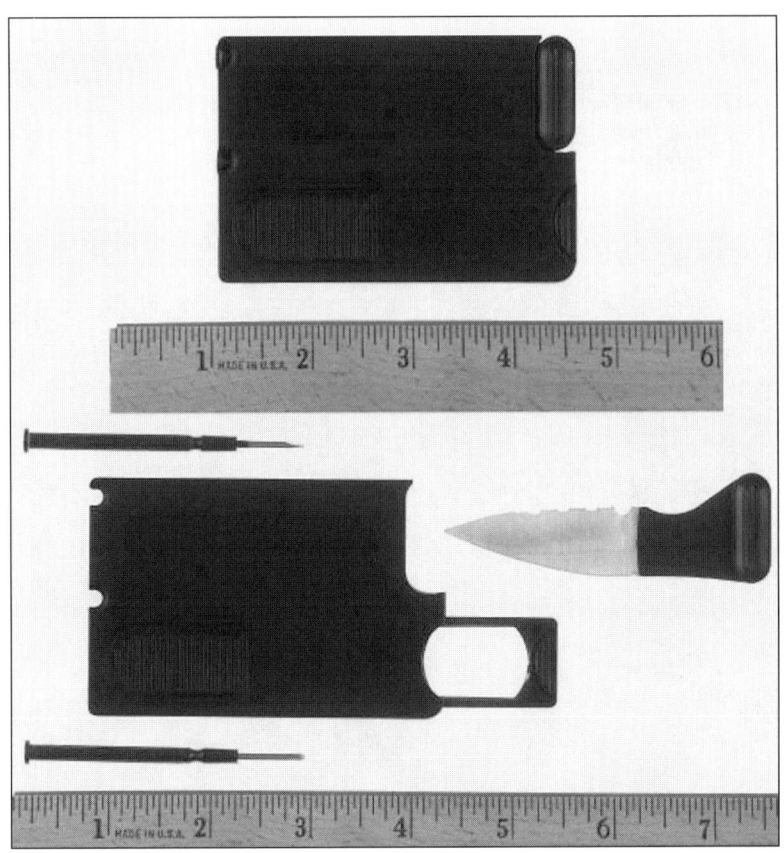

## TOOL CARD

Manufacturer: Unknown
Composition: Plastic case with metal knife blade
Note: Designed to be the approximate size of a credit card so it can be carried in a pocket. Also includes magnifying glass, two screwdrivers, and a flashlight.

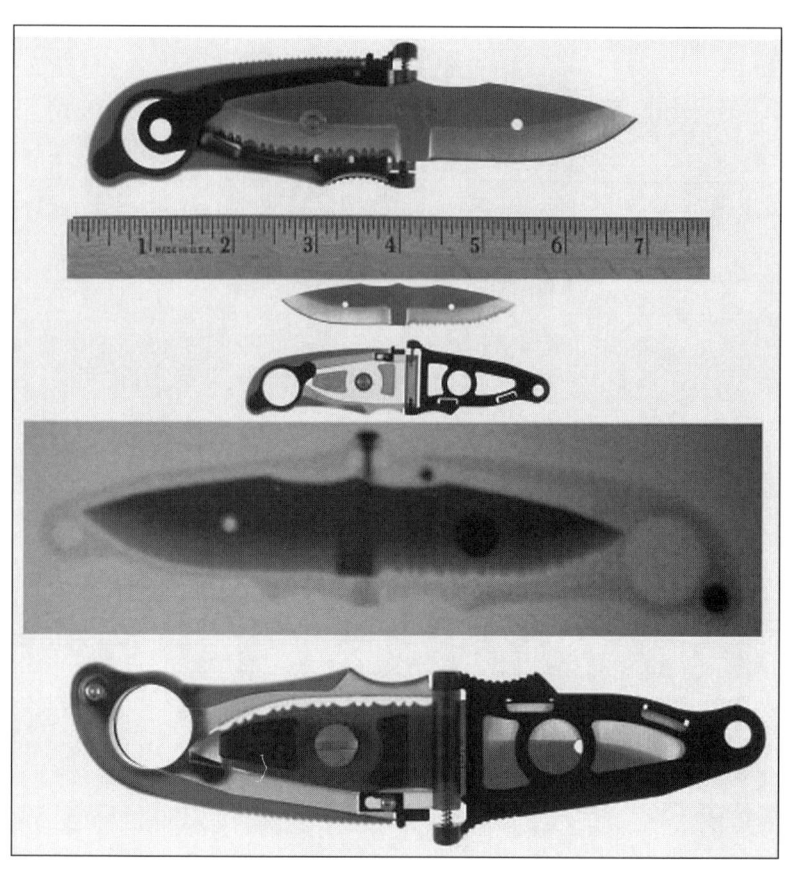

## DUO KNIFE

Manufacturer: SOG
Made in Japan
Composition: Reversible metal blade with plastic handle
Note: Plastic handle folds to expose either plain or serrated edge of removable knife blade.

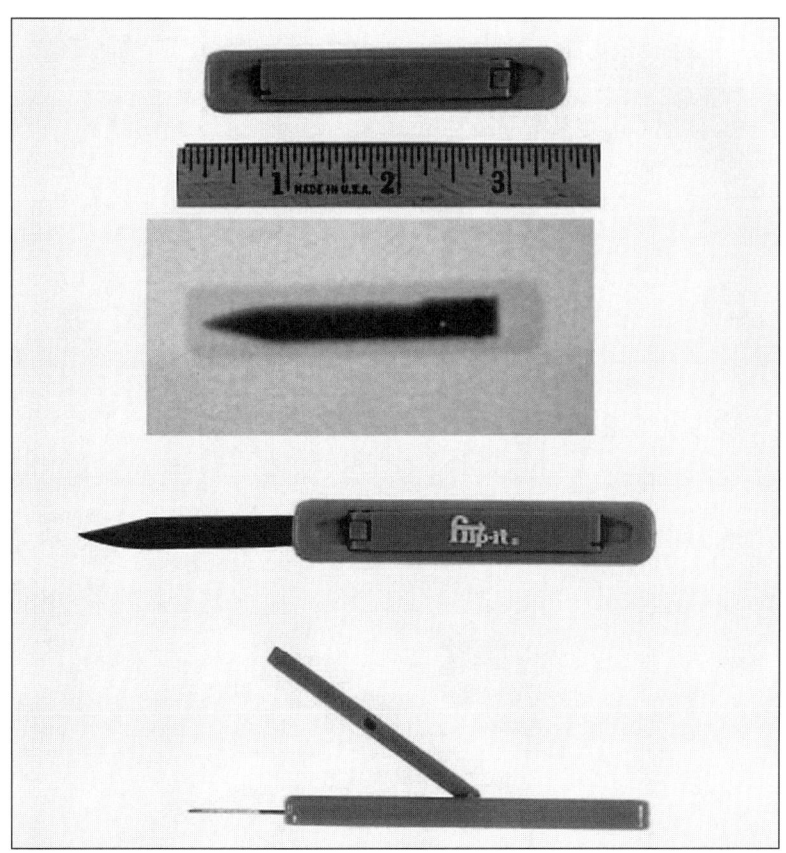

## FLIP KNIFE

Manufacturer: Flip-It
Made in USA
Composition: Plastic and metal
Note: Knife blade is deployed by lifting the handle insert, sliding it forward, and flipping it over to lock the blade in place.

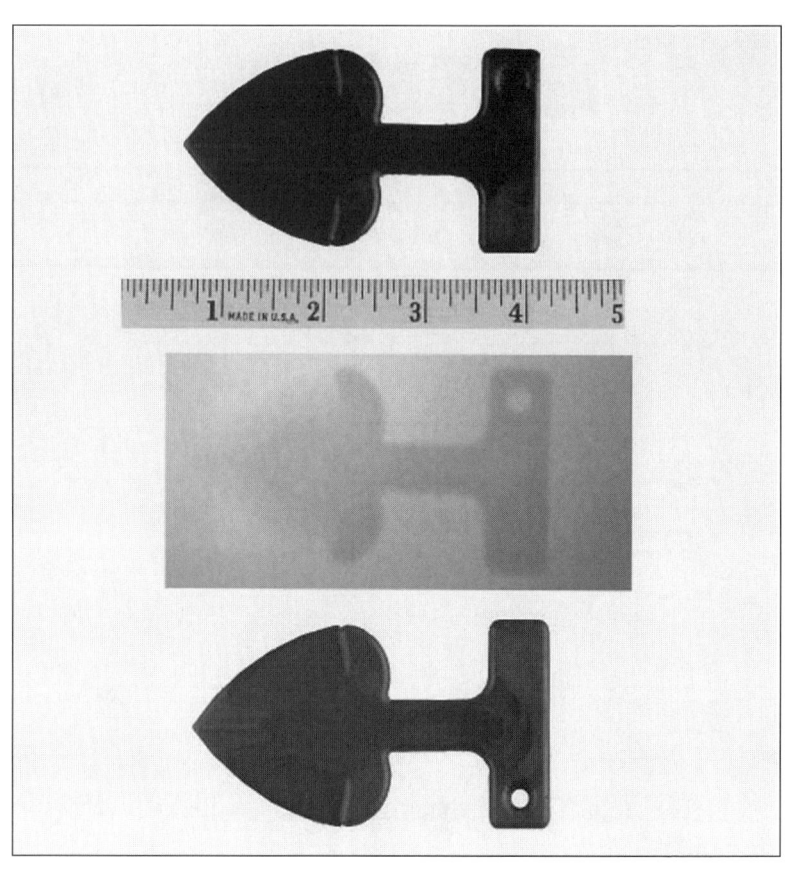

## PUSH DAGGER

Manufacturer: Unknown
Made in ?
Composition: Plastic
Note: All-plastic construction makes this invisible to a magnetometer.

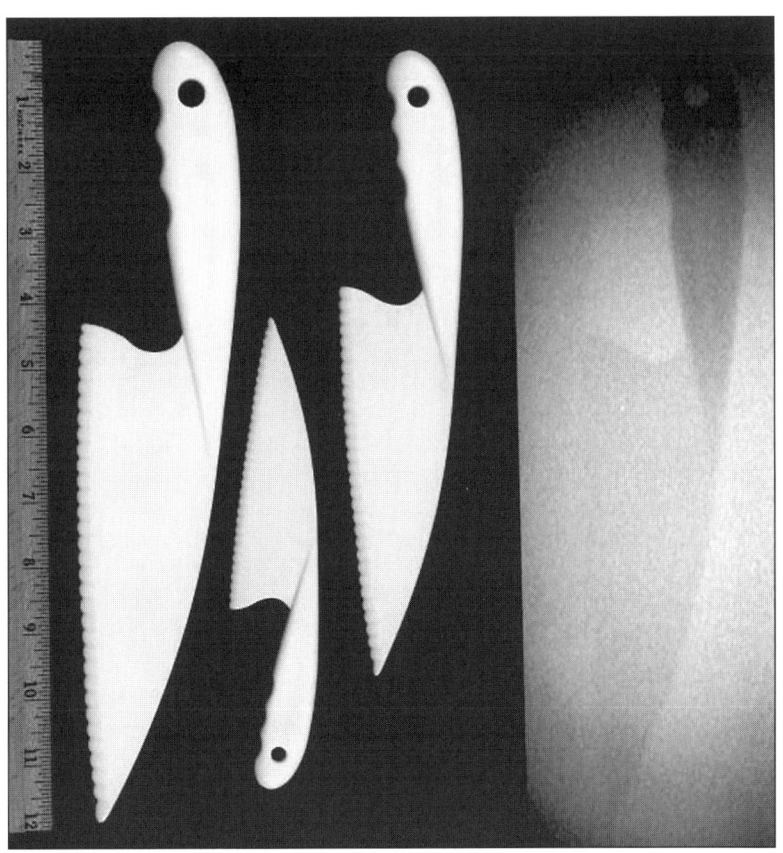

## LETTUCE KNIVES

Manufacturer: Unknown
Made in ?
Composition: Plastic
Note: Inexpensive set of three various-sized knives that are invisible to a magnetometer.

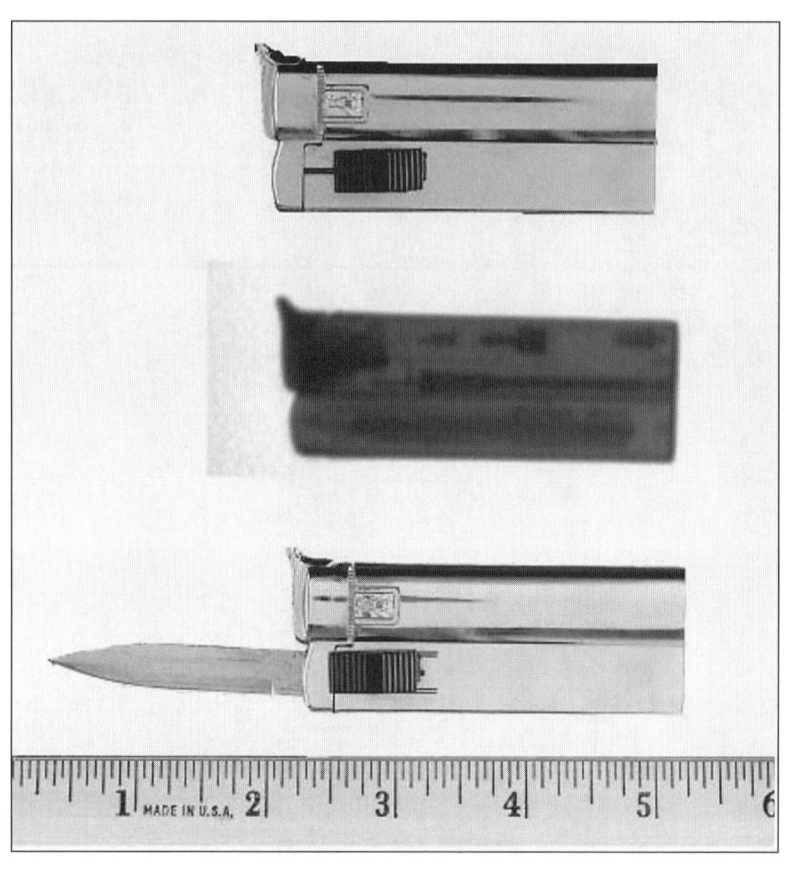

## LIGHTER KNIFE

Manufacturer: Unknown
Made in ?
Composition: Metal
Note: Working lighter that deploys a knife blade by activation of a button on the side.

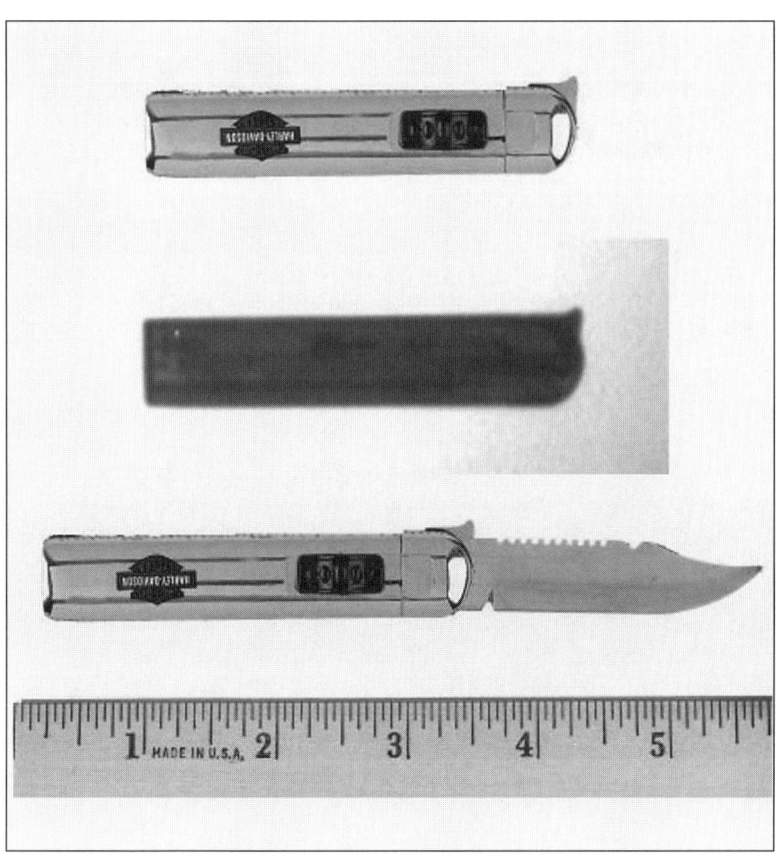

## LIGHTER KNIFE

Manufacturer: Unknown
Made in ?
Composition: Metal
Note: Working lighter that deploys a knife blade by activation of a button on the side.

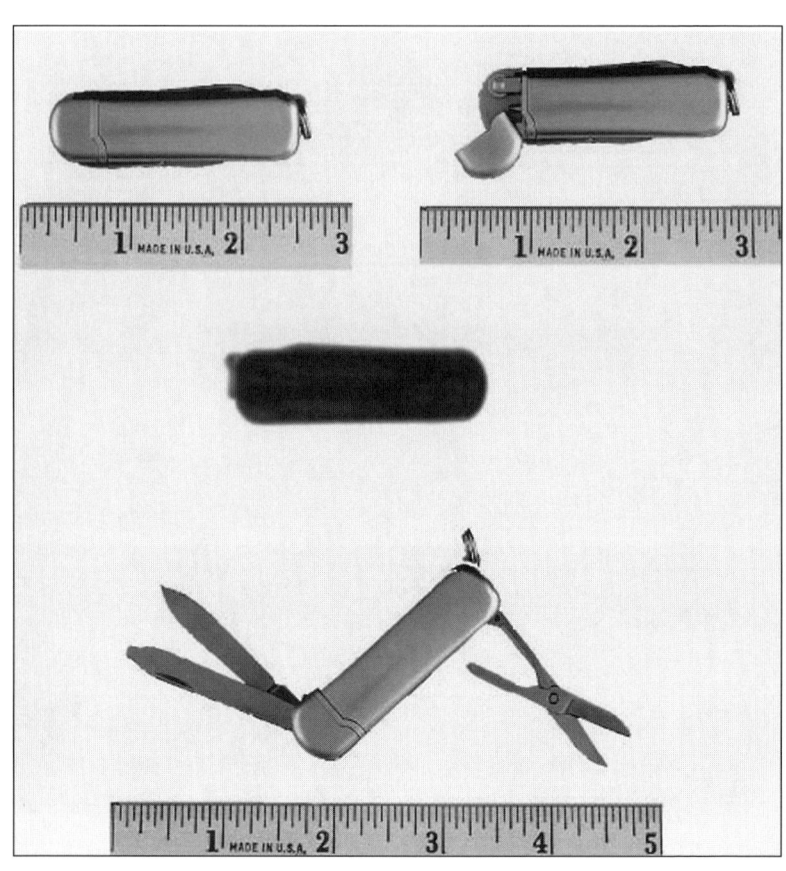

## LIGHTER KNIFE

Manufacturer: Howell
Made in Korea
Composition: Metal
Note: Working lighter that stores a knife blade, nail file, and scissors.

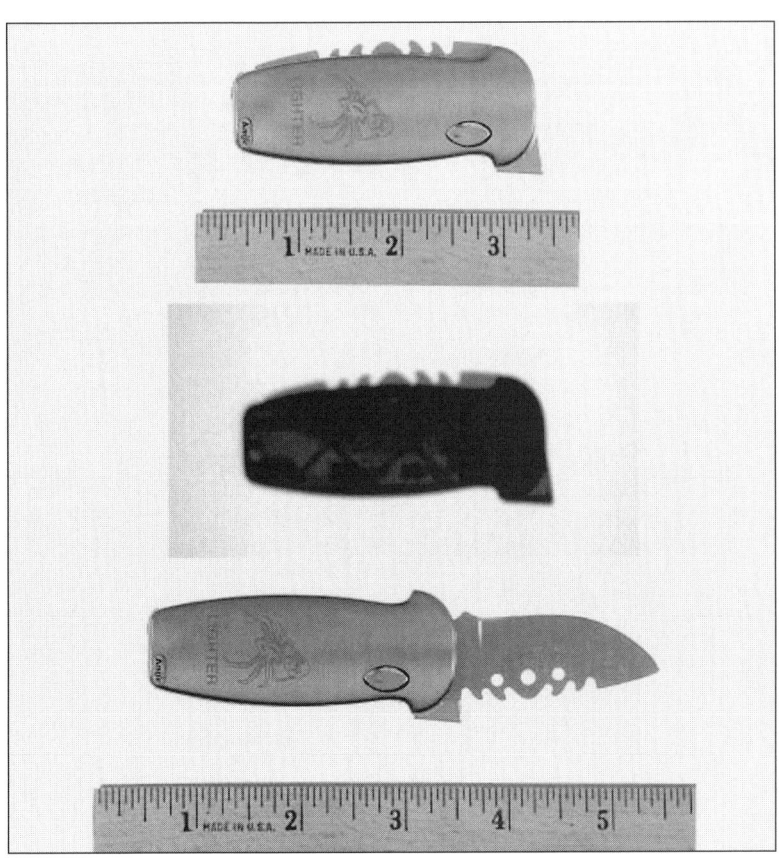

**CIGARETTE LIGHTER KNIFE**

Manufacturer: Unknown
Made in ?
Composition: Metal knife blade in metal case handle
Note: This working lighter conceals an automatic knife.

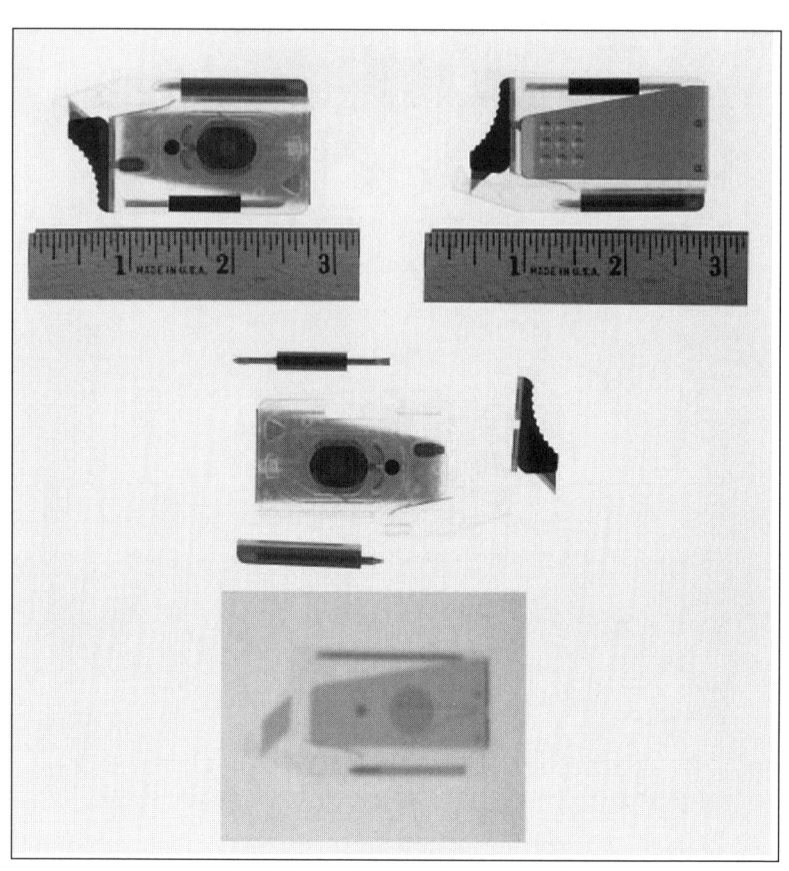

## TOOL CARD/MONEY CLIP

Manufacturer: Unknown
Made in USA
Composition: Plastic and metal
Note: This money clip stores a blade, two screwdrivers, and a flashlight.

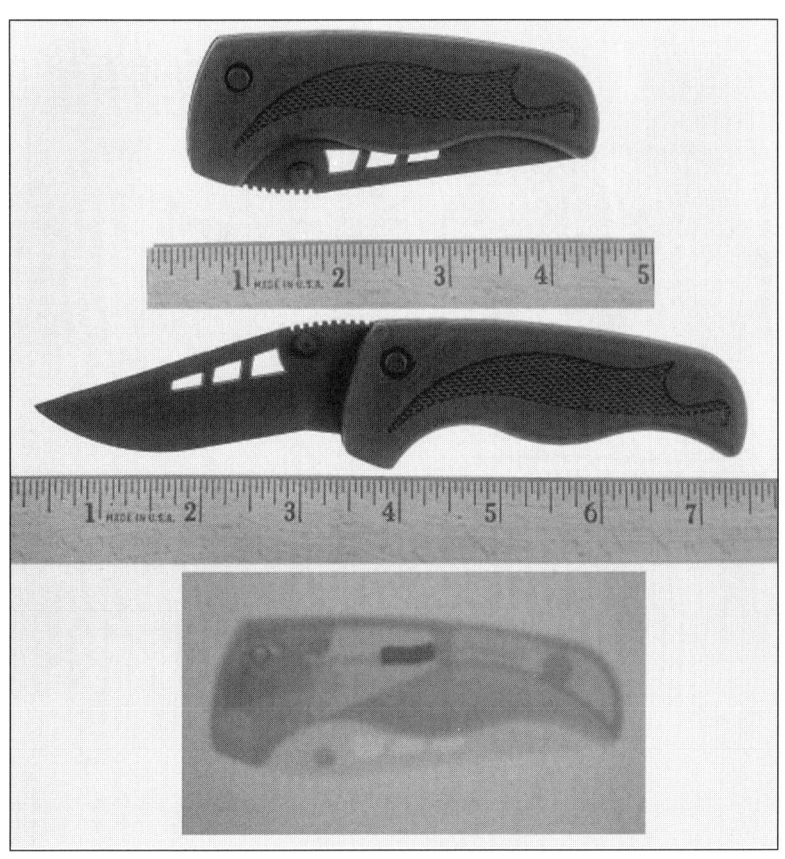

## PLASTIC FOLDING KNIFE

Manufacturer: Meyerco
Made in USA
Composition: Plastic and metal
Note: The knife is all-plastic construction except for a small spring to assist in opening.

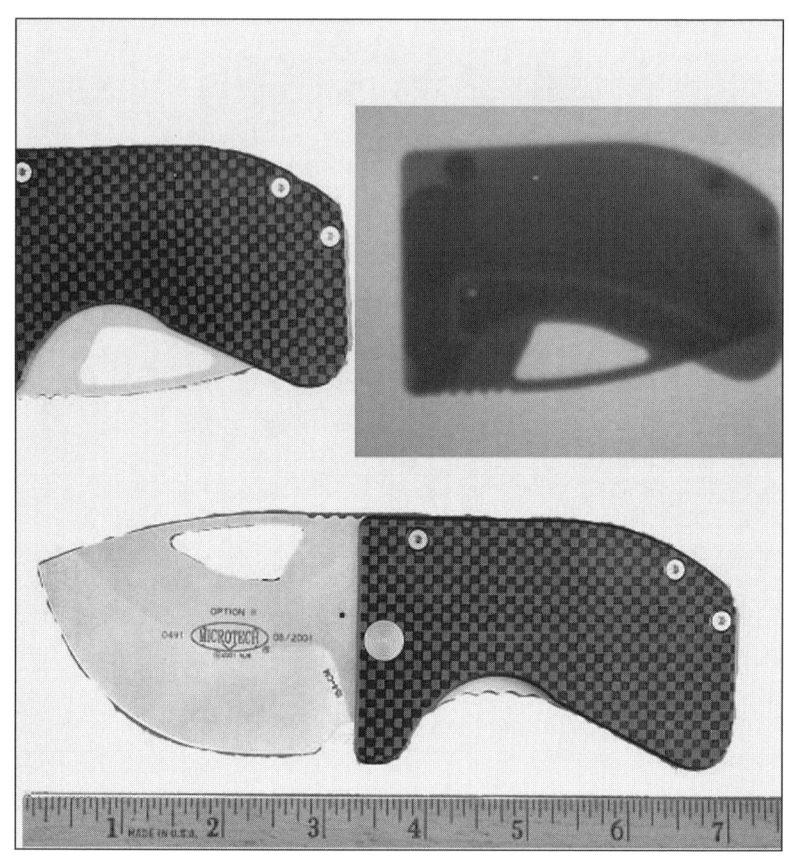

## CARD KNIFE

Manufacturer: Microtech (model name—Option II)
Made in USA
Composition: Plastic and metal

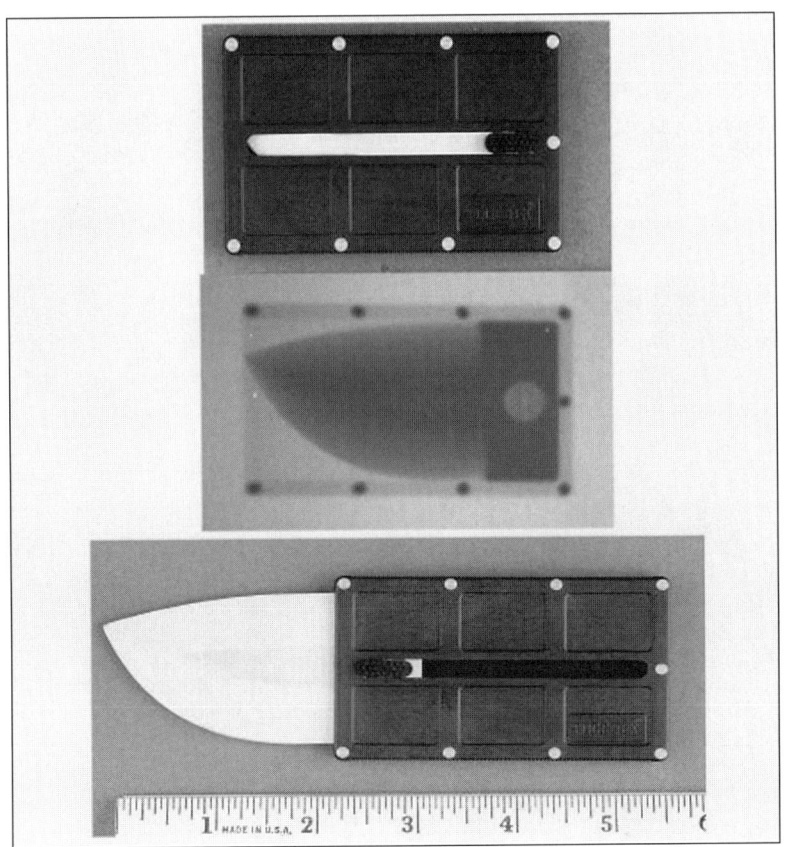

## CARD KNIFE

Manufacturer: Edge-Tek
Made in USA
Composition: Plastic and metal
Note: Blade slides out of the handle.

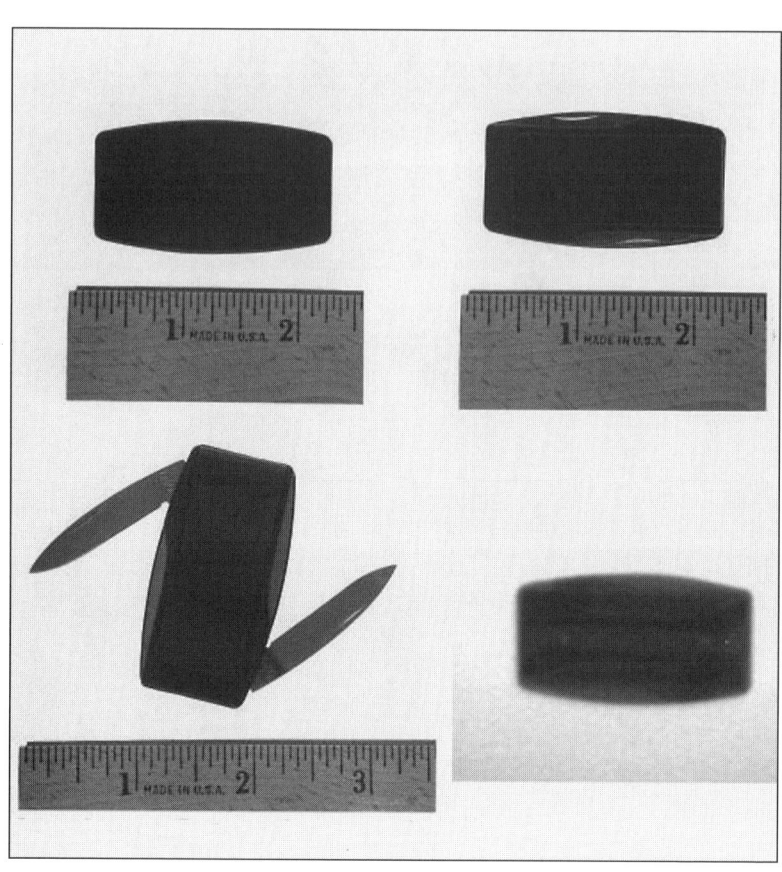

## MONEY CLIP KNIFE

Manufacturer: Zippo
Made in USA
Composition: Metal
Note: Money clip houses a knife blade and nail file.

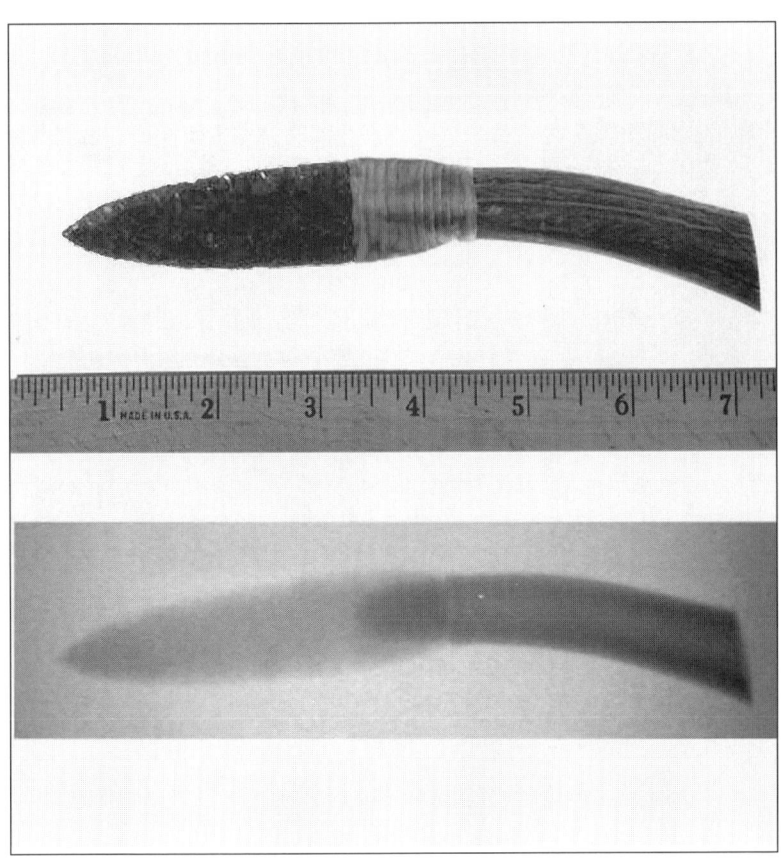

## OBSIDIAN KNIFE

Manufacturer: These knives are custom made.
Composition: Obsidian blade with a bone handle
Note: No metal is used in this knife, making it invisible to a magnetometer.

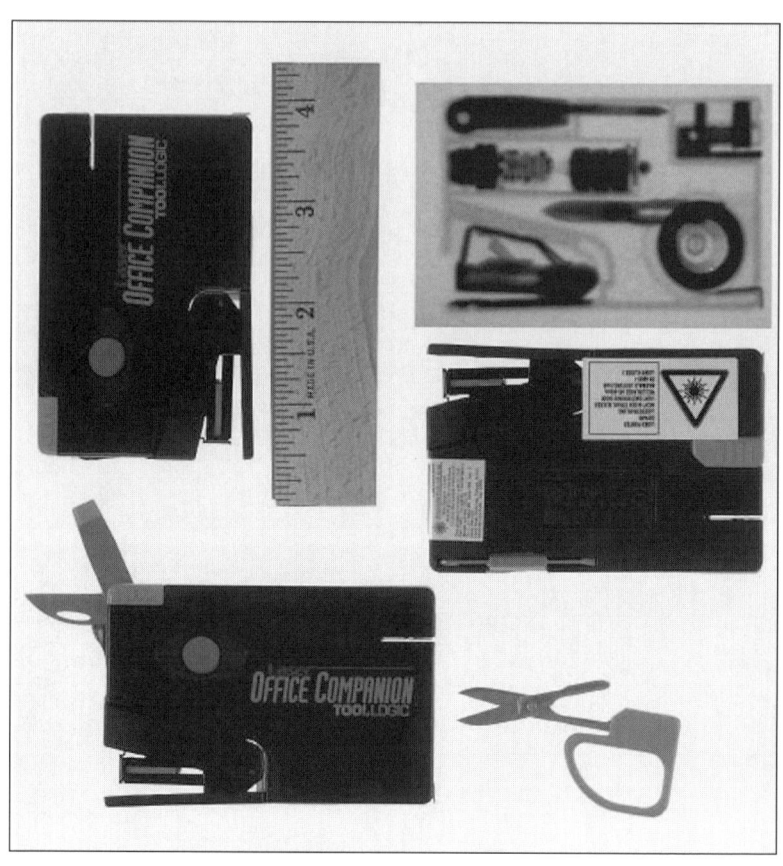

## OFFICE TOOL KIT

Manufacturer: Tool Logic (model name—Office Companion)
Made in USA
Composition: Plastic and metal
Note: The length and width of a credit card, it contains a stapler, ballpoint pen, and other common office tools.

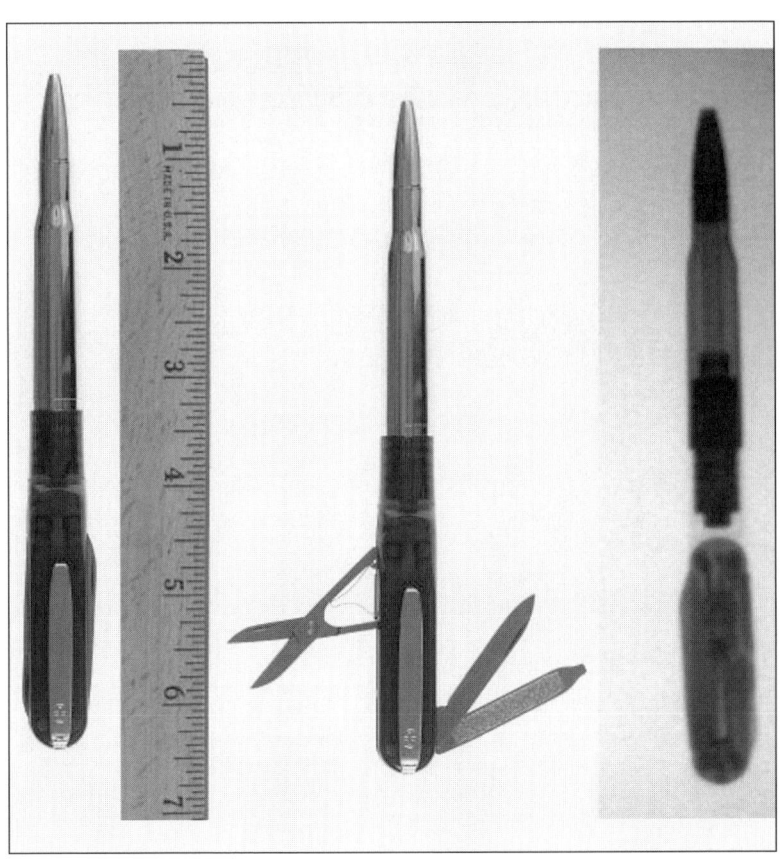

## INK PEN KNIFE

Manufacturer: Victorinox
Made in Switzerland
Composition: Metal blade with plastic handle
Note: Pen looks like a rifle cartridge. There is a knife blade, nail file, scissors, and a flashlight stored in the pen handle.

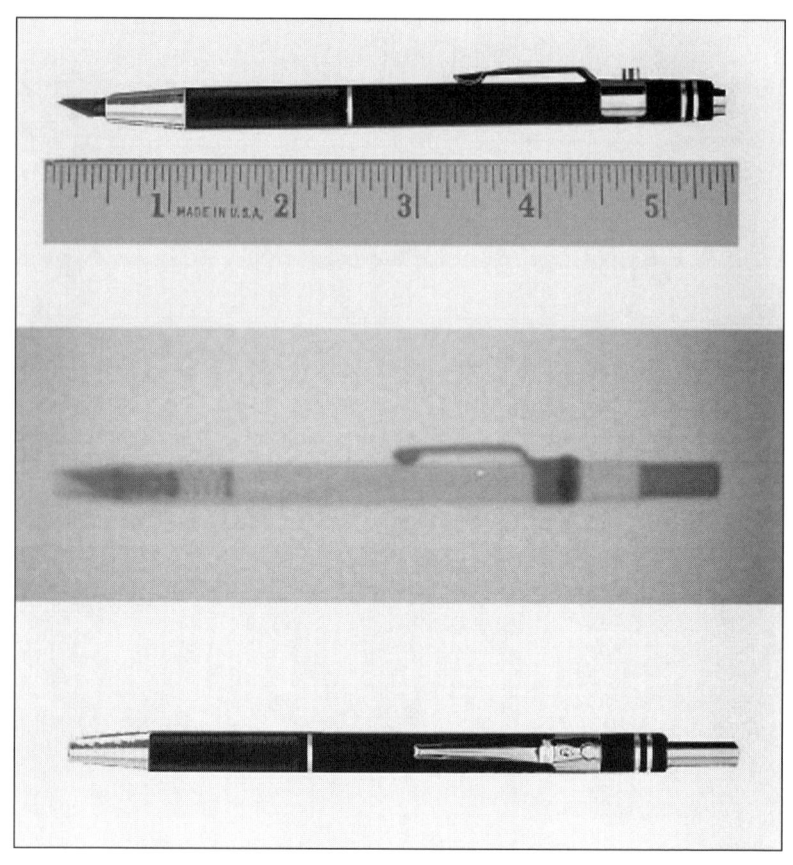

## RAZOR KNIFE

Manufacturer: X-ACTO (model name—X-Calibre)
Made in ?
Composition: All metal
Note: Looks like an ink pen. Push button deploys blade.

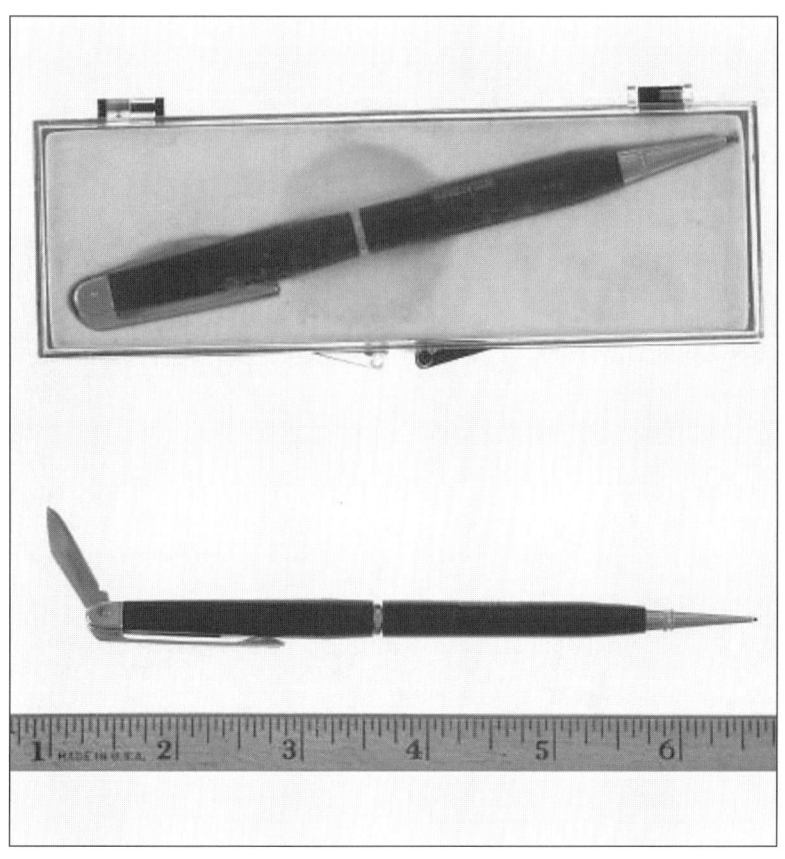

## INK PEN KNIFE

Manufacturer: Unknown
Made in USA
Composition: Metal
Note: Looks like an ink pen. Knife blade folds in the side of the pen.

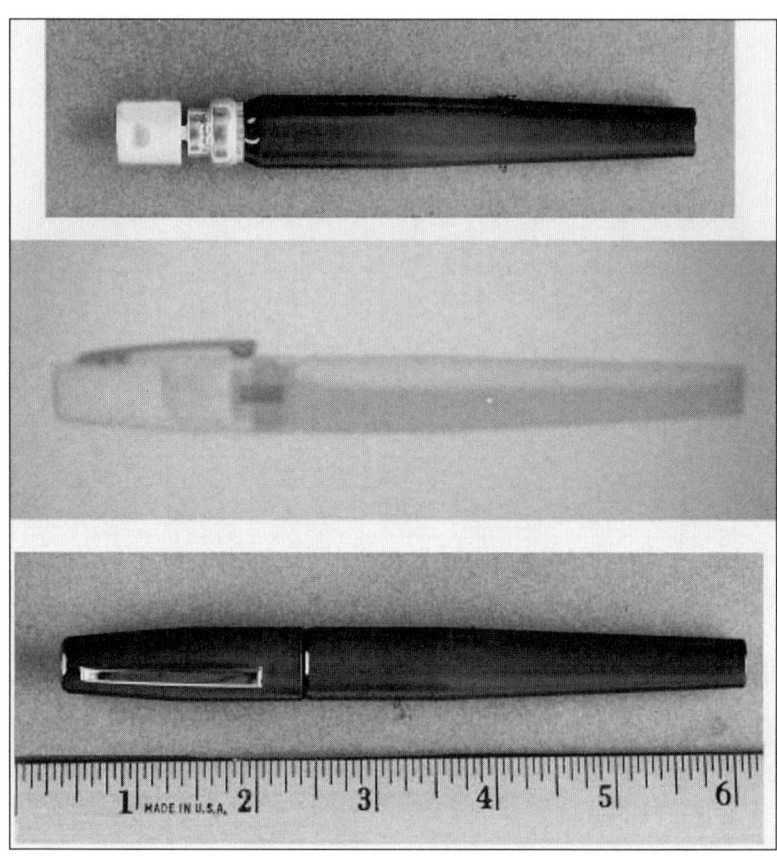

## INK PEN PEPPER SPRAY

Manufacturer: Unknown
Made in USA
Composition: Plastic and metal
Note: Looks like an ink pen and can eject pepper spray.

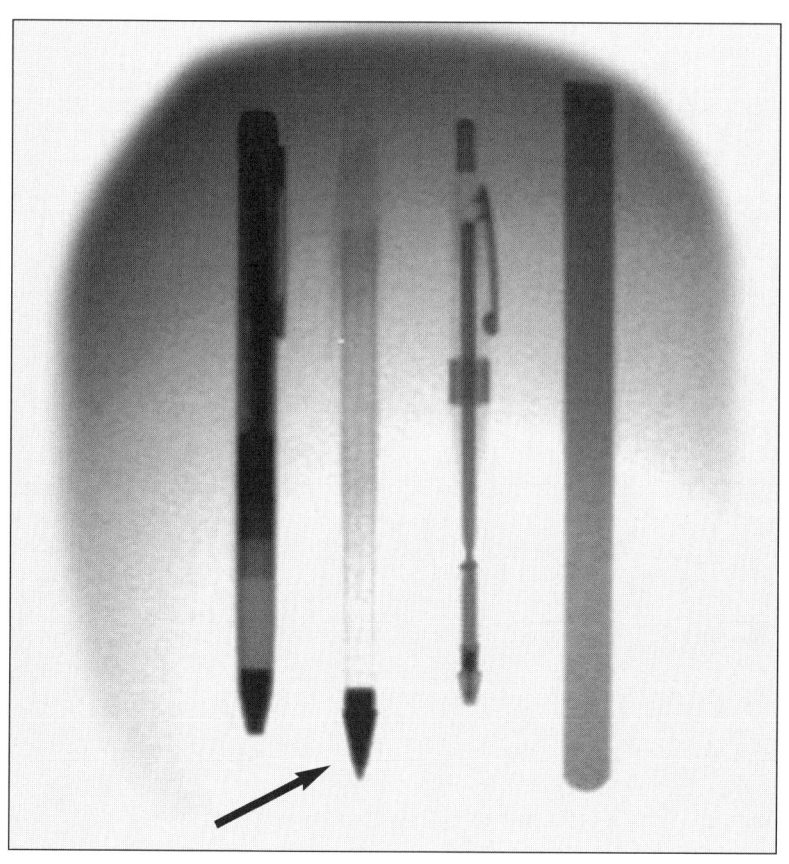

## PENKNIFE

X-ray of a penknife with other pens; the penknife is the second image from the left.

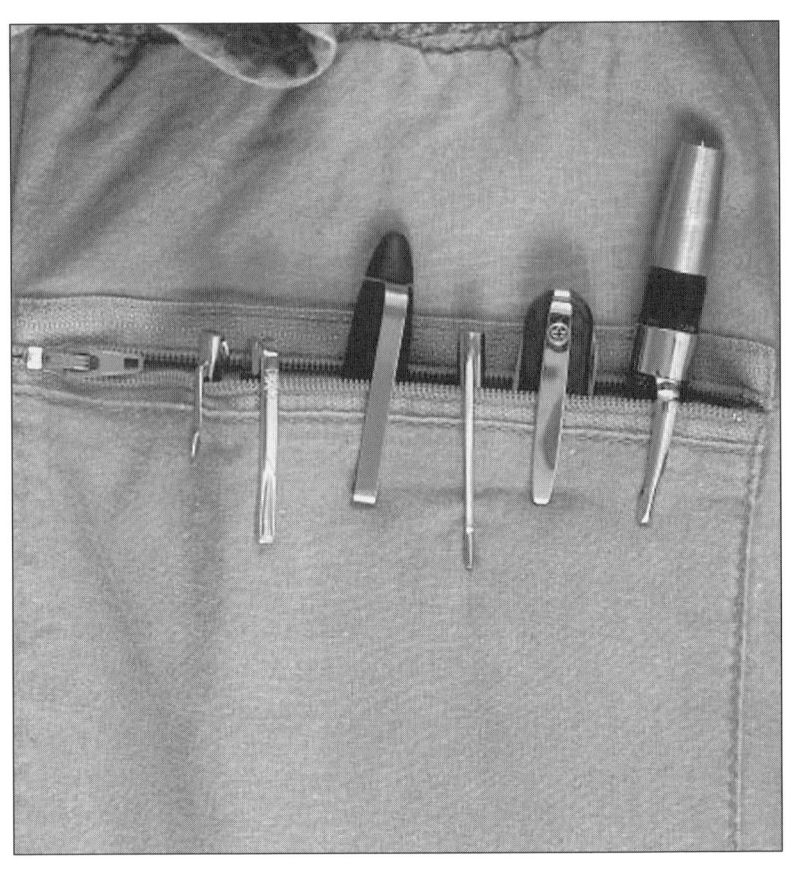

**PEN-TYPE KNIVES**

This photograph shows the profile of the various penknives when carried in a pocket.

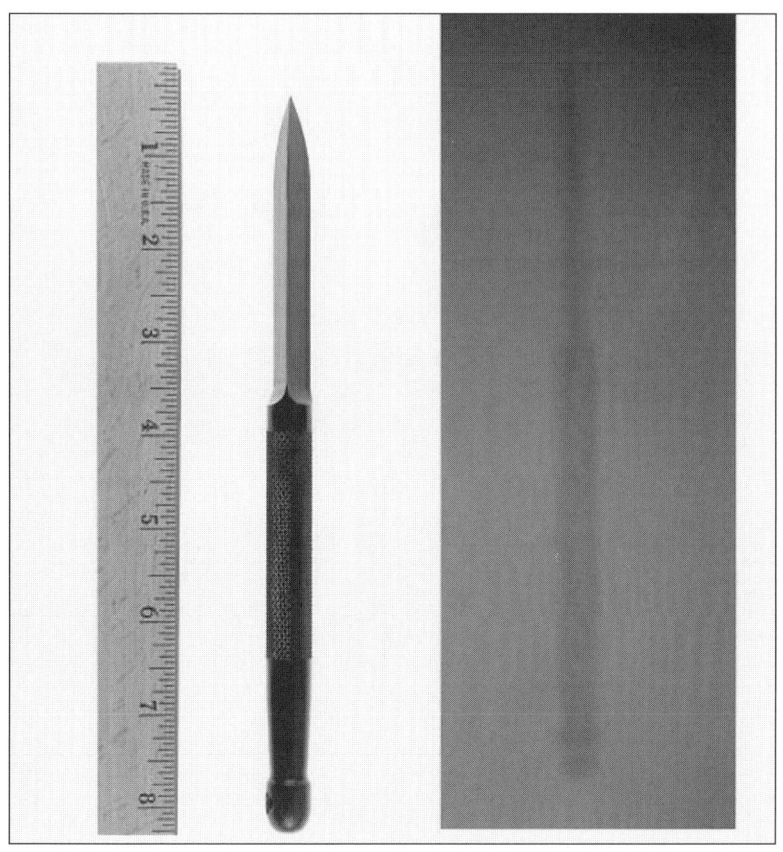

## PLASTIC SPIKE

Manufacturer: Unknown
Made in ?
Composition: Polymer/plastic
Note: All-plastic construction of the spike (dagger) makes it invisible to a magnetometer.

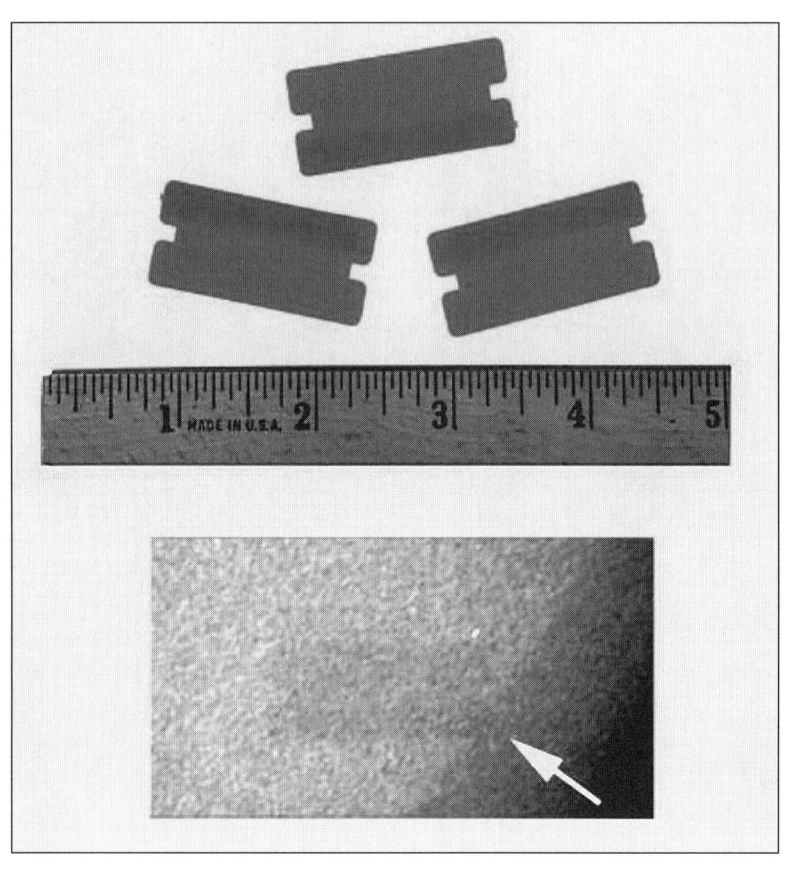

## PLASTIC RAZOR BLADES

Manufacturer: Unknown
Made in China
Composition: Plastic
Note: These are the same size as standard metal razor blades. All-plastic construction makes them invisible to a magnetometer.

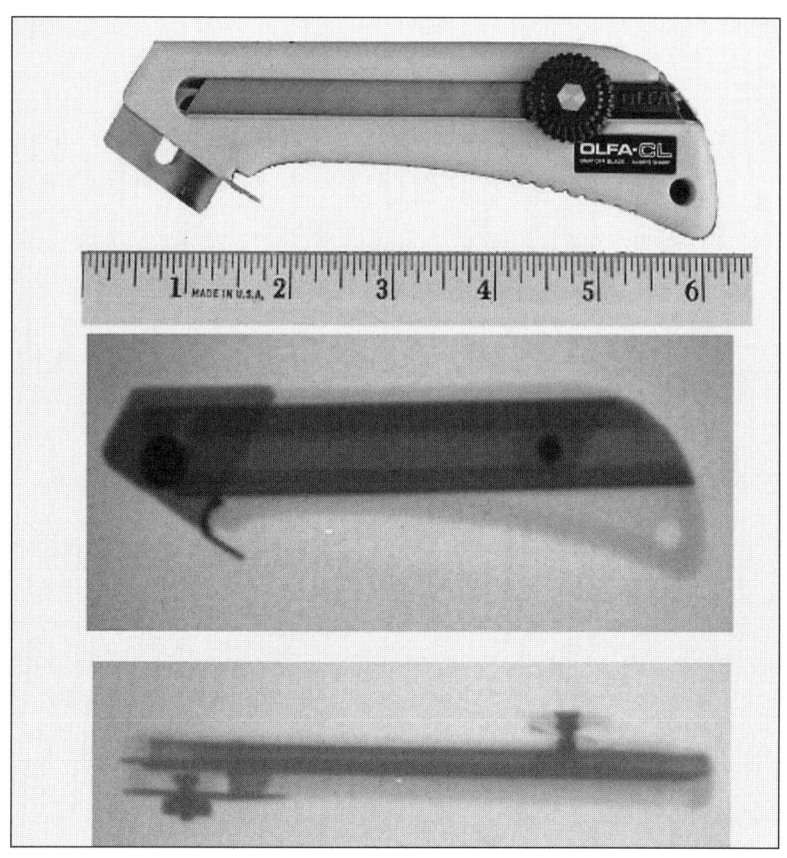

**BOX CUTTER KNIFE**

Manufacturer: OLFA
Made in Japan
Composition: Metal razor blade with plastic handle
Note: Razor blade slides out of handle.

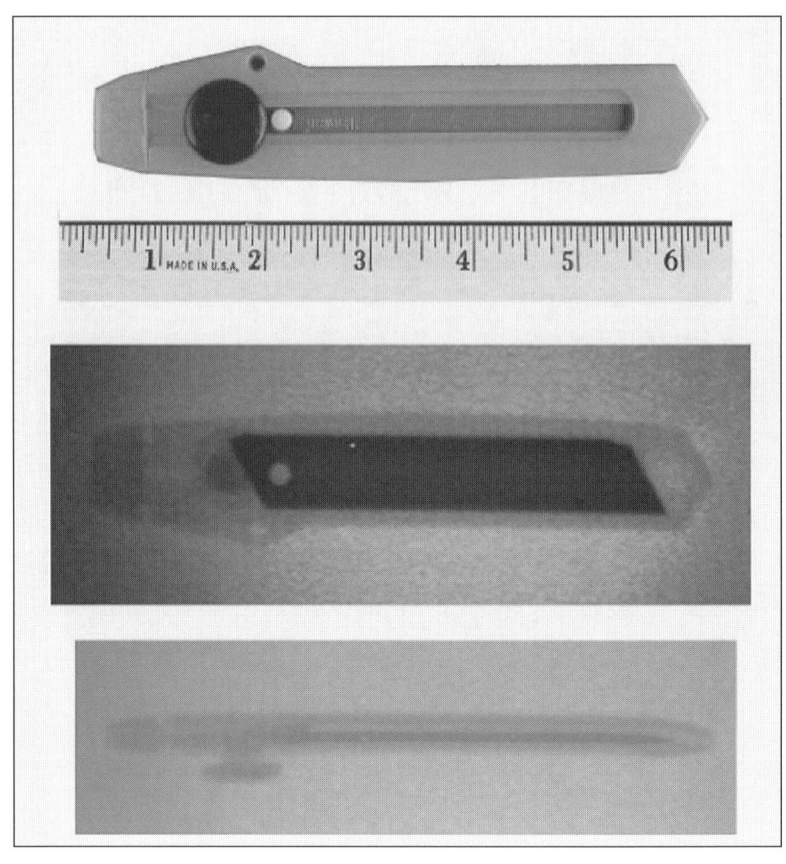

## RAZOR KNIFE

Manufacturer: Unknown
Made in Taiwan
Composition: Metal razor blade with plastic handle
Note: Razor blade slides out of handle.

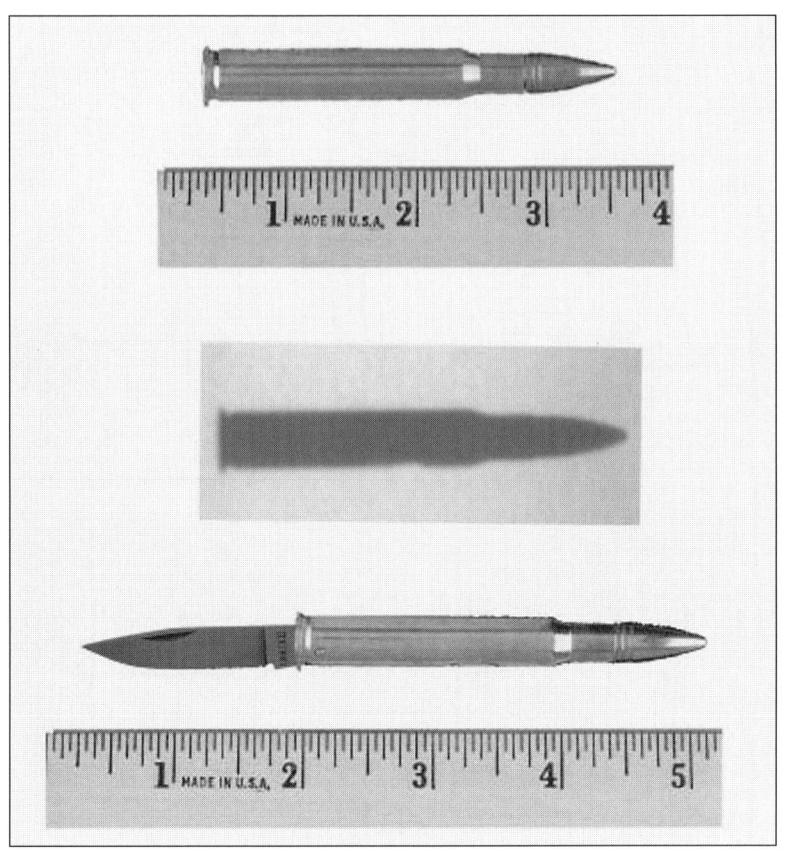

## BULLET (CARTRIDGE) KNIFE

Manufacturer: United Cutlery
Made in Italy
Composition: Metal blade inside of metal cartridge
Note: Looks like a rifle cartridge. Headstamp reads "30-06 SPRG."

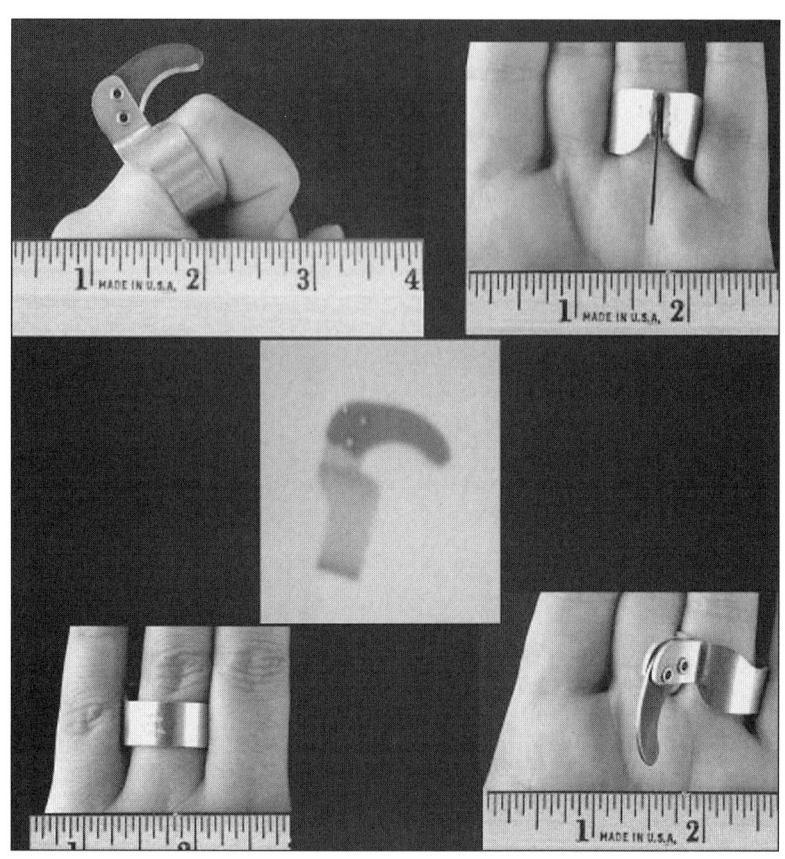

**RING KNIFE**

Manufacturer: Bates
Made in USA
Composition: Metal blade with ring
Note: Blade attached to a ring. Commonly used to cut strings on packages and hay bales.

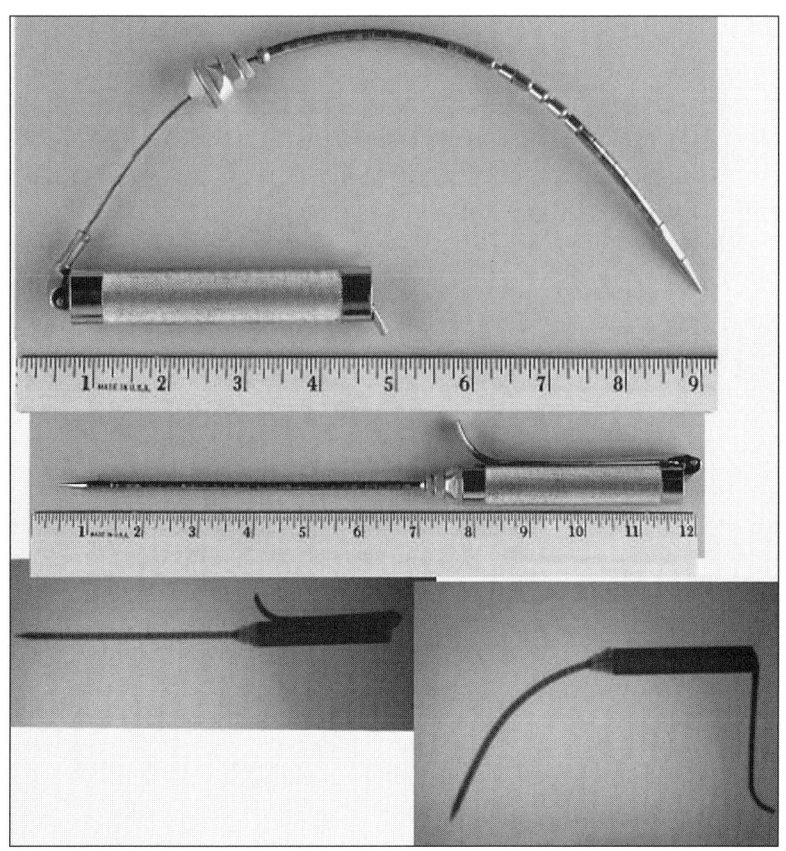

## SHUCKRA

Manufacturer: Unknown
Made in ?
Composition: Metal
Note: By locking the handle and pulling on an internal wire, this device becomes taut and may be used as a spike.

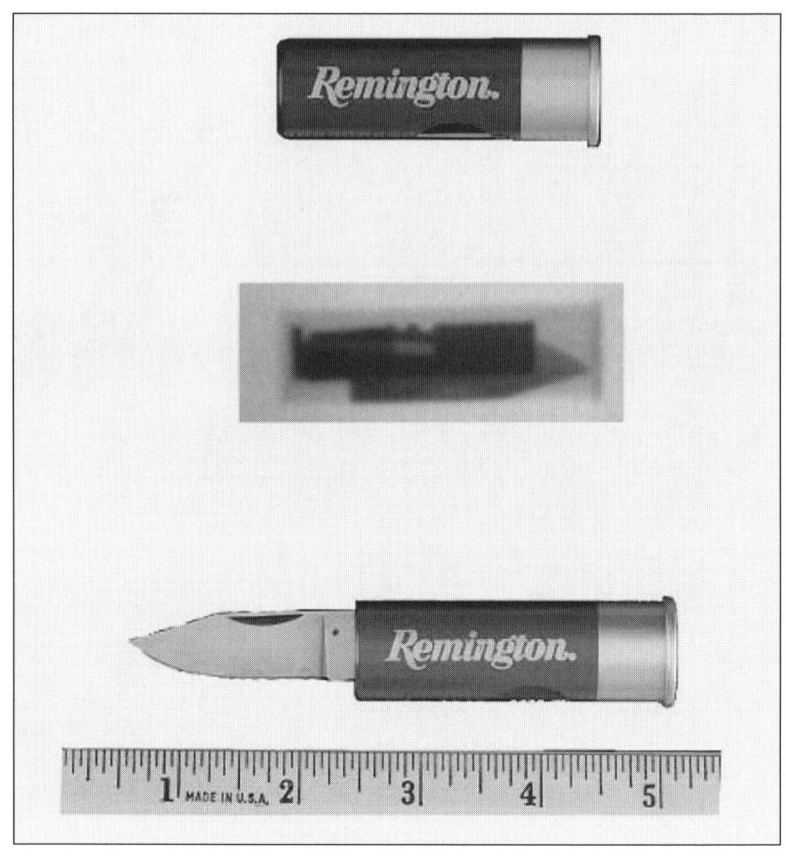

## SHOTSHELL KNIFE

Manufacturer: United Cutlery
Made in Italy
Composition: Metal blade inside plastic shotshell
Note: Looks like a shotshell for a shotgun. Headstamp reads "12-GA REMINGTON PETERS."

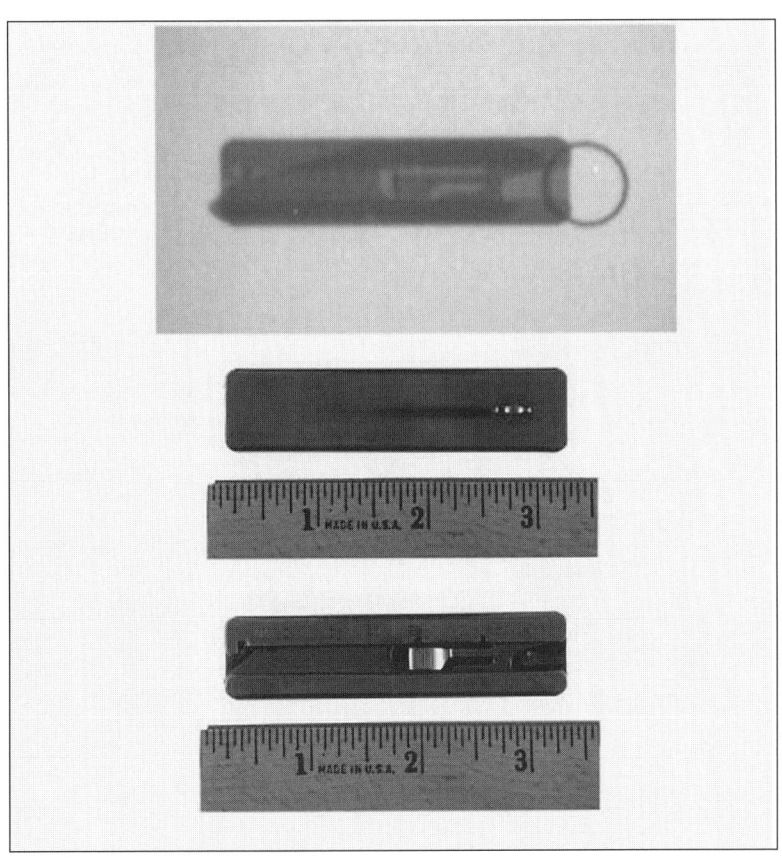

## SLIDER KNIFE

Manufacturer: Unknown
Made in USA
Composition: Metal razor blade with metal handle
Note: Handle looks like a ruler. The blade has a lock button that allows the knife to be deployed.

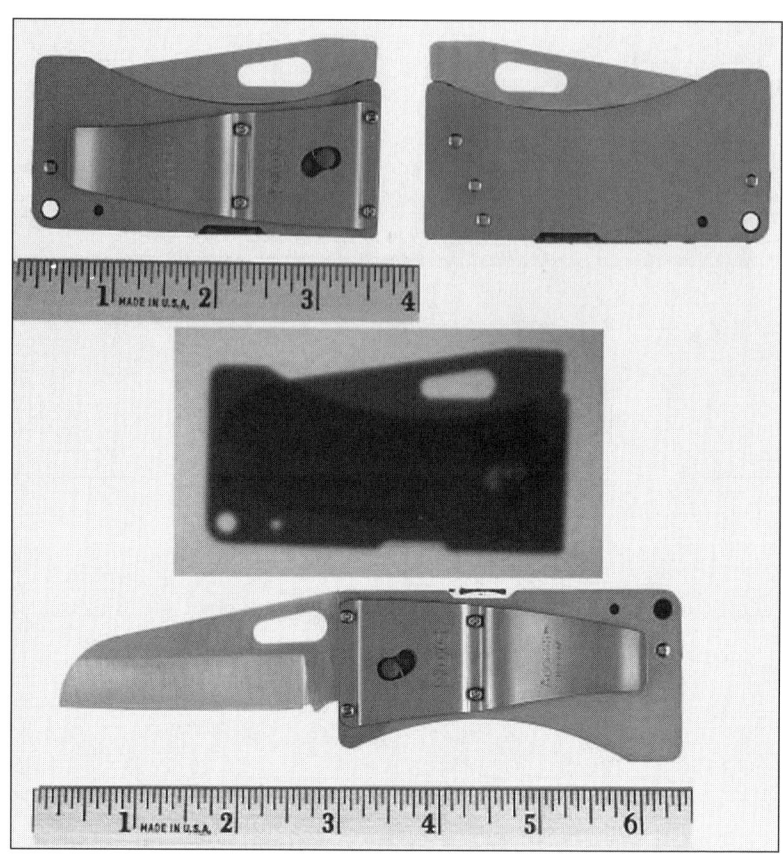

## SOG ACCESS CARD

Manufacturer: SOG
Made in Japan
Composition: Metal blade and handle
Note: This knife houses tweezers, a toothpick, and a straight pin.

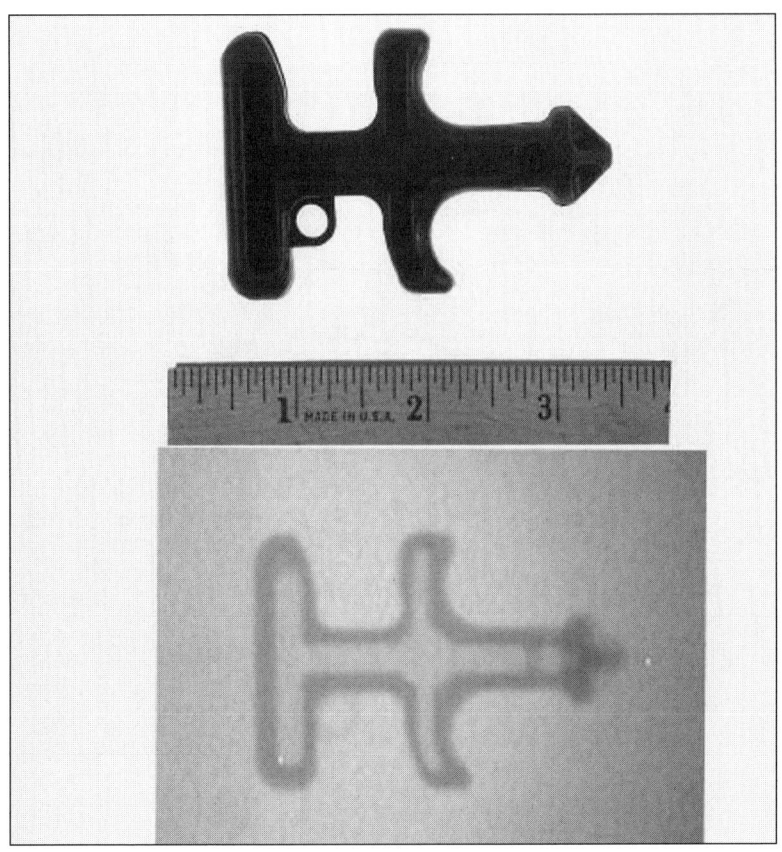

## STINGER KEY CHAIN

Manufacturer: Shomer-Tec
Made in USA
Composition: All-polymer/plastic striking tool
Note: Comes in black, blue, or clear. All-plastic construction makes it invisible to a magnetometer.

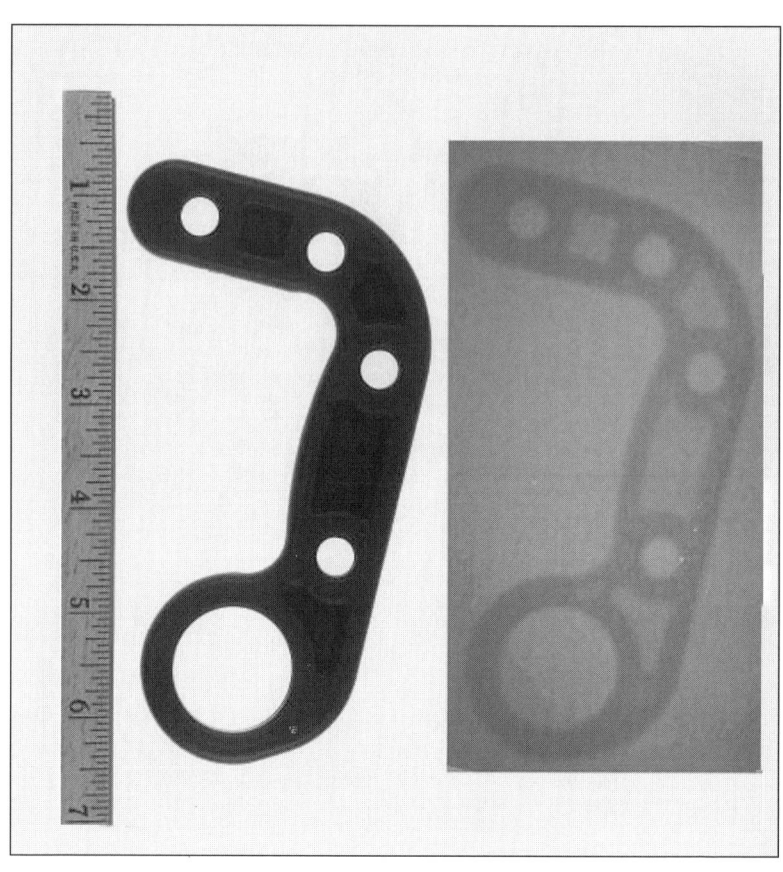

## IMPACT KERAMBIT

Manufacturer: Shomer-Tec
Made in USA
Composition: All-polymer/plastic
Note: This impact tool is made of plastic, making it invisible to a magnetometer.

## THROWING CARDS

Manufacturer: Unknown
Made in ?
Composition: Metal
Note: These all-metal cards are the size of standard playing cards. They are edged on all sides and designed to be thrown at a target.

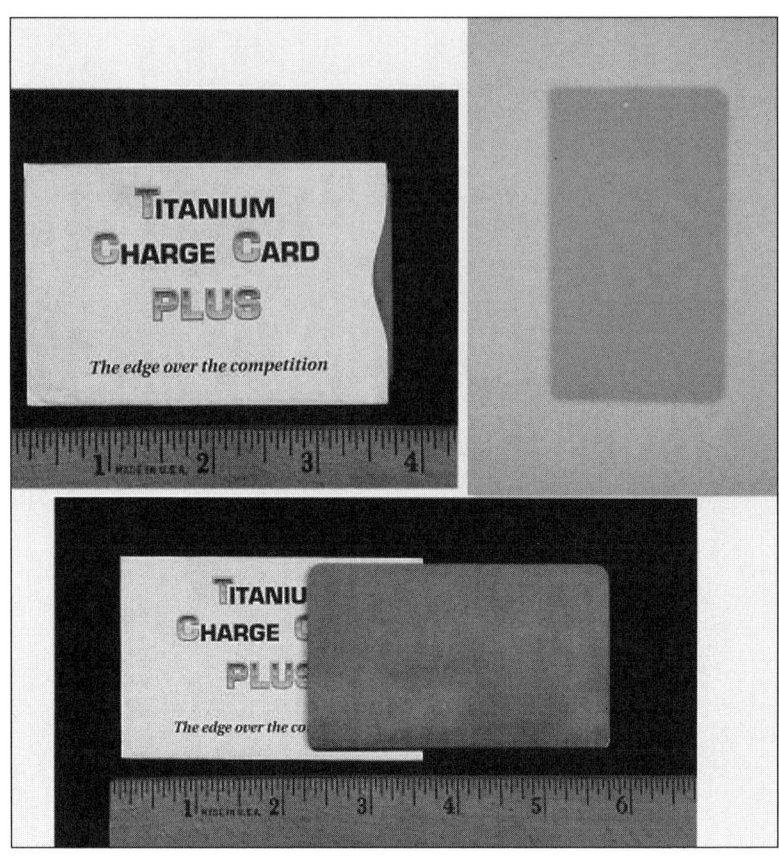

## TITANIUM CHARGE CARD PLUS

Manufacturer: Shomer-Tec
Made in USA
Composition: Metal
Note: Edged on one side, this card is designed to fit in your wallet like a credit card.

## MONEY CLIP

Manufacturer: Kershaw
Made in Japan
Composition: Metal
Note: Money clip conceals a blade, nail file, and scissors.

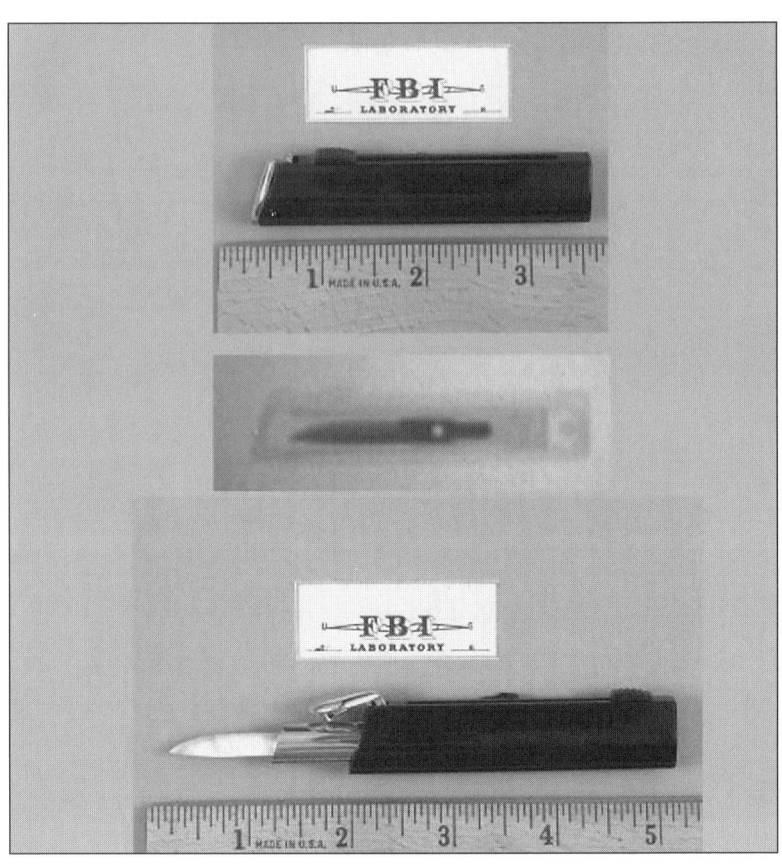

## LIPSTICK

Manufacturer: United Cutlery
Made in Taiwan
Composition: Plastic case with metal band
Note: Appears to be a cosmetic. Blade extends by sliding the button along the side of the case.

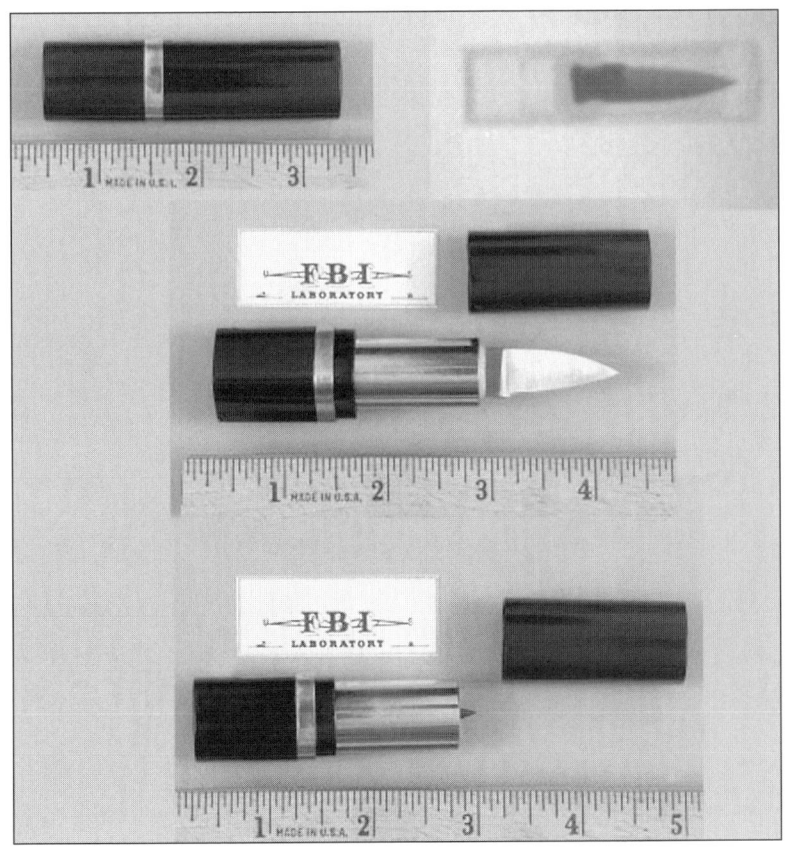

## LIPSTICK

Manufacturer: United Cutlery
Made in Taiwan
Composition: Plastic case with metal band
Note: Appears to be a cosmetic. Blade extends by twisting the case.

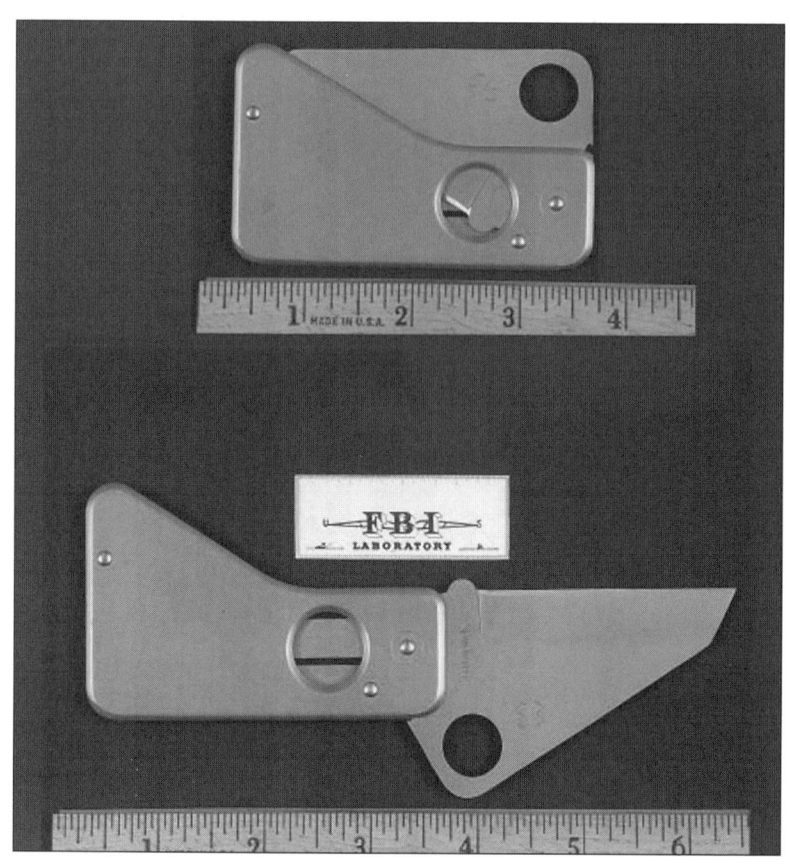

## CARD KNIFE

Manufacturer: Spyderco
Made in Japan
Composition: Metal
Note: Many companies make credit card-sized knives. Styles vary by manufacturer.

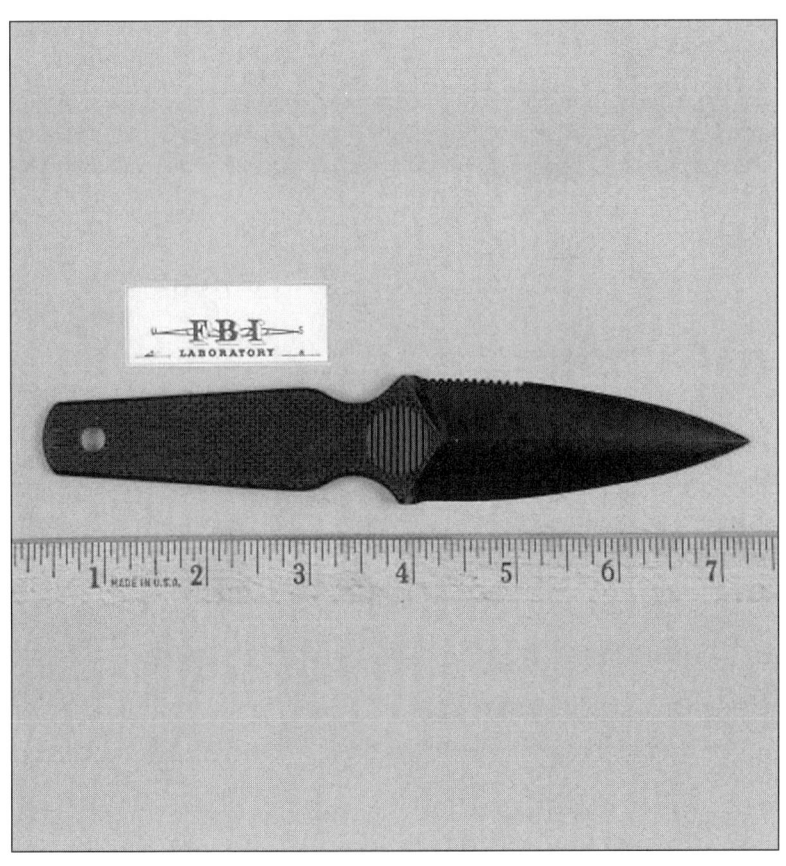

## PLASTIC KNIFE

Manufacturer: Lansky
Made in USA
Composition: Plastic
Note: Other companies make knives similar to this one. The style of knife may vary by manufacturer.

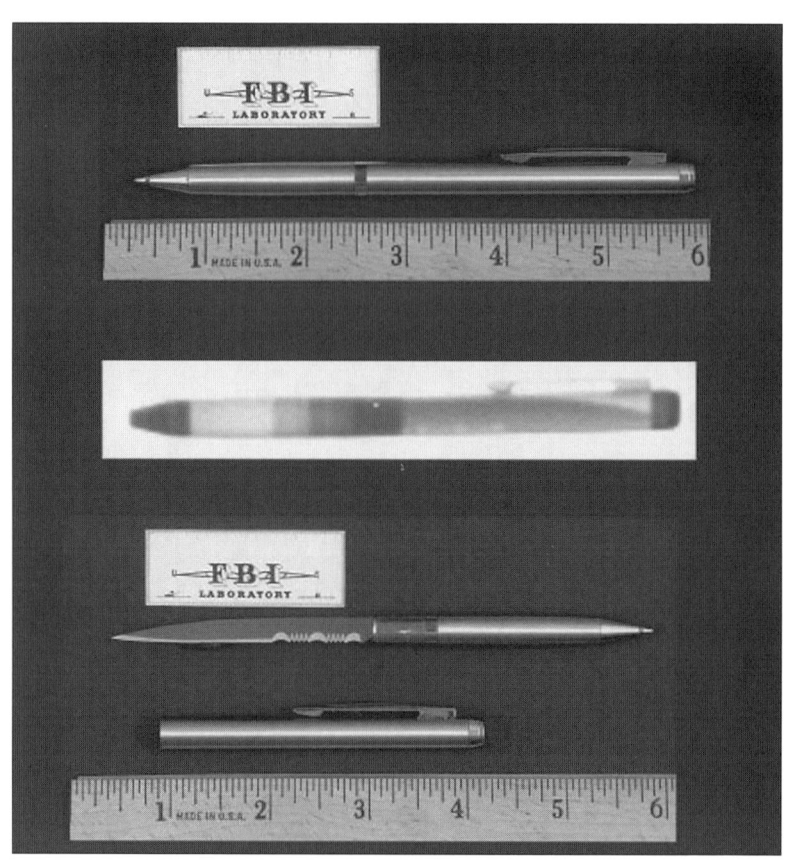

## PENKNIFE

Manufacturer: United Cutlery
Made in Taiwan
Composition: Metal
Note: Styles vary to include different blade styles and case colors.

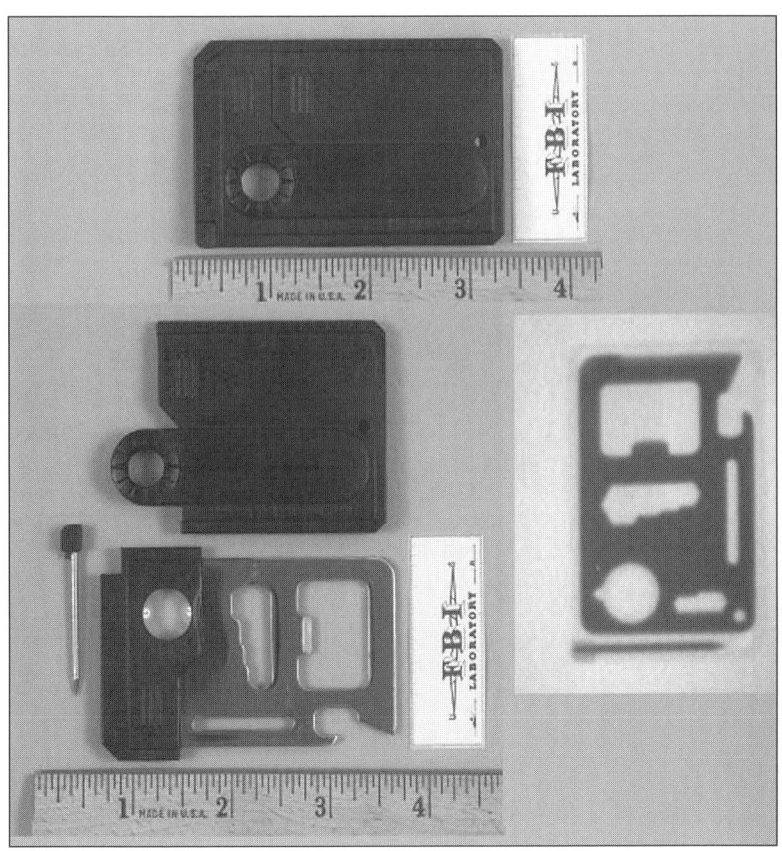

## CREDIT CARD-STYLE UTILI-TOOL

Manufacturer: Unknown
Made in Taiwan
Composition: Plastic case with metal blade
Note: Includes a can opener, screwdriver, bottle opener, wrenches, and a magnifying glass.

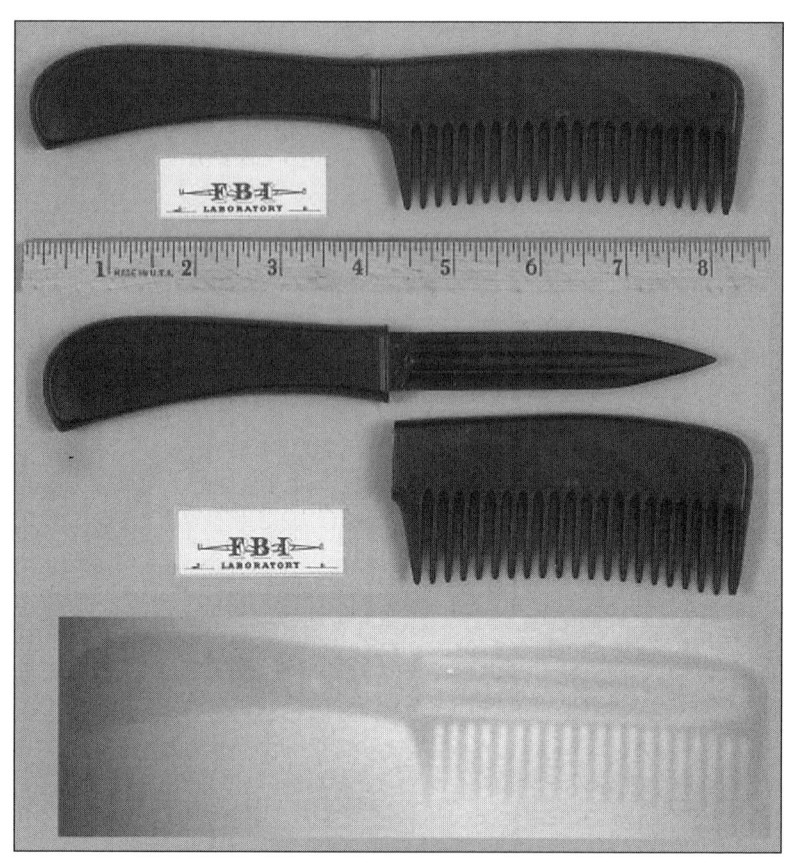

## COMB

Manufacturer: United Cutlery
Made in Taiwan
Composition: Plastic
Note: Appears to be a normal comb. Handle pulls from head, which acts as a sheath.

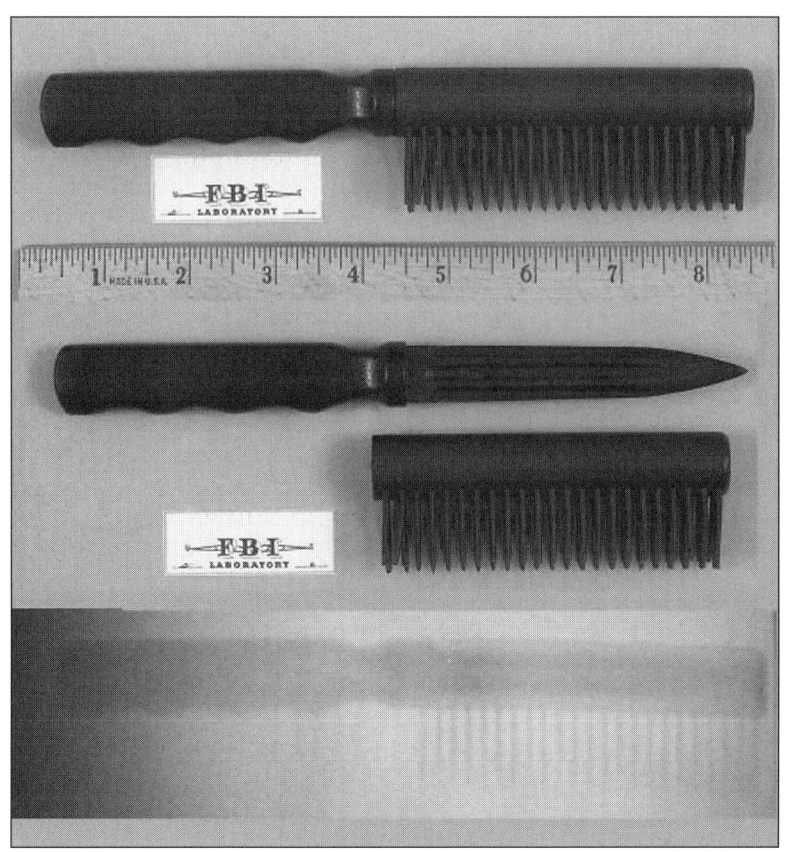

## HAIRBRUSH

Manufacturer: United Cutlery
Made in Taiwan
Composition: Plastic
Note: Appears to be a normal hairbrush. Handle pulls from head, which acts as a sheath.

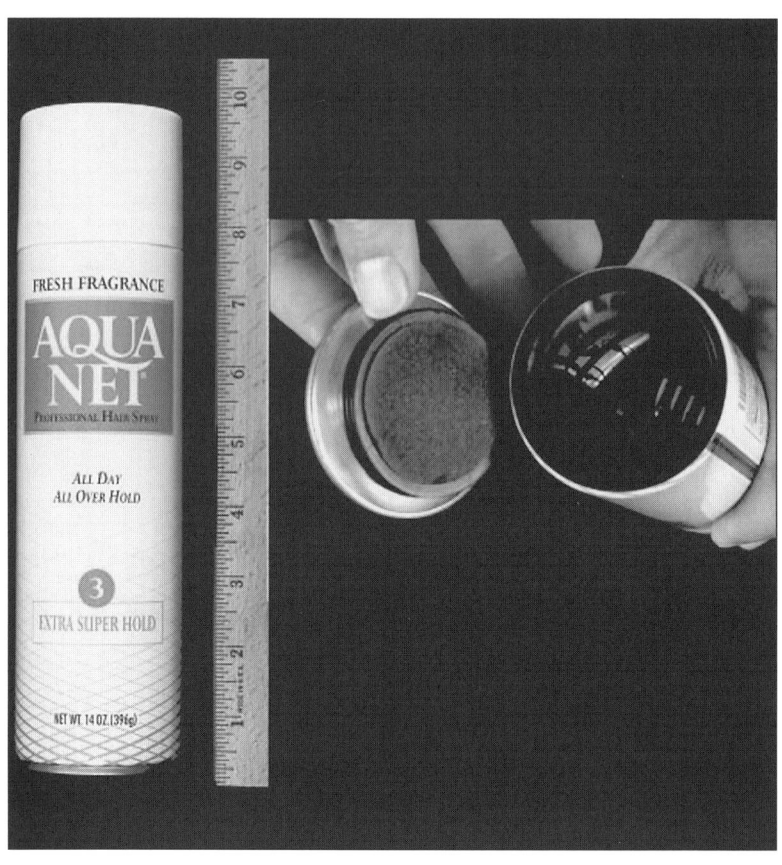

## CAN SAFES

Can safes open at the bottom so that items may be hidden inside.

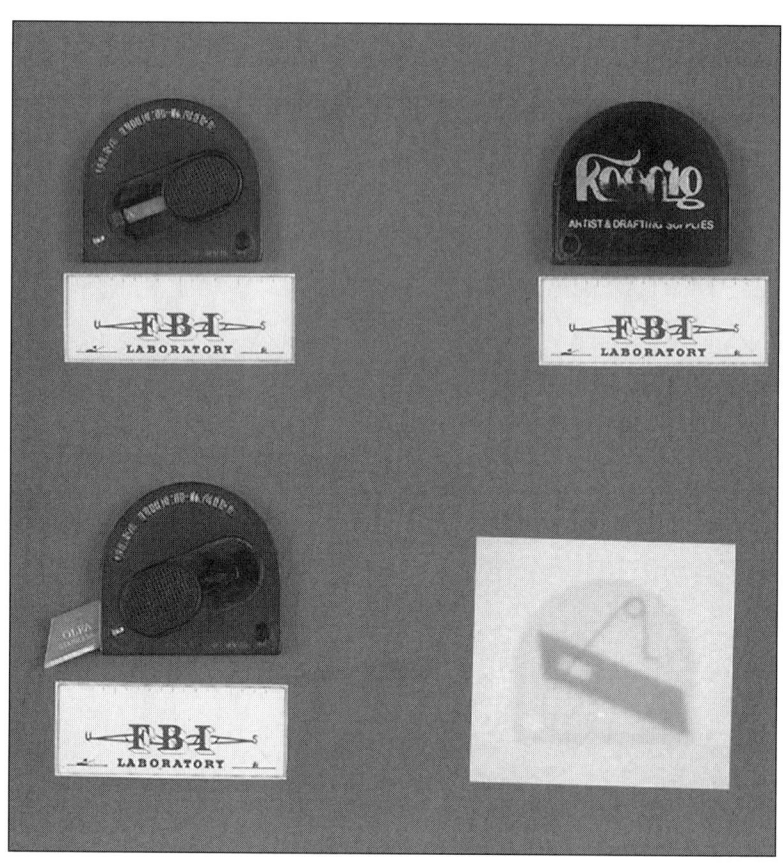

## BOX CUTTER KNIFE

Manufacturer: OLFA
Made in Japan
Composition: Plastic case with metal blade
Note: Colors and styles vary.

## BOOK SAFE

This is an example of a book safe. The pages are one solid piece and are hollowed out on the inside.

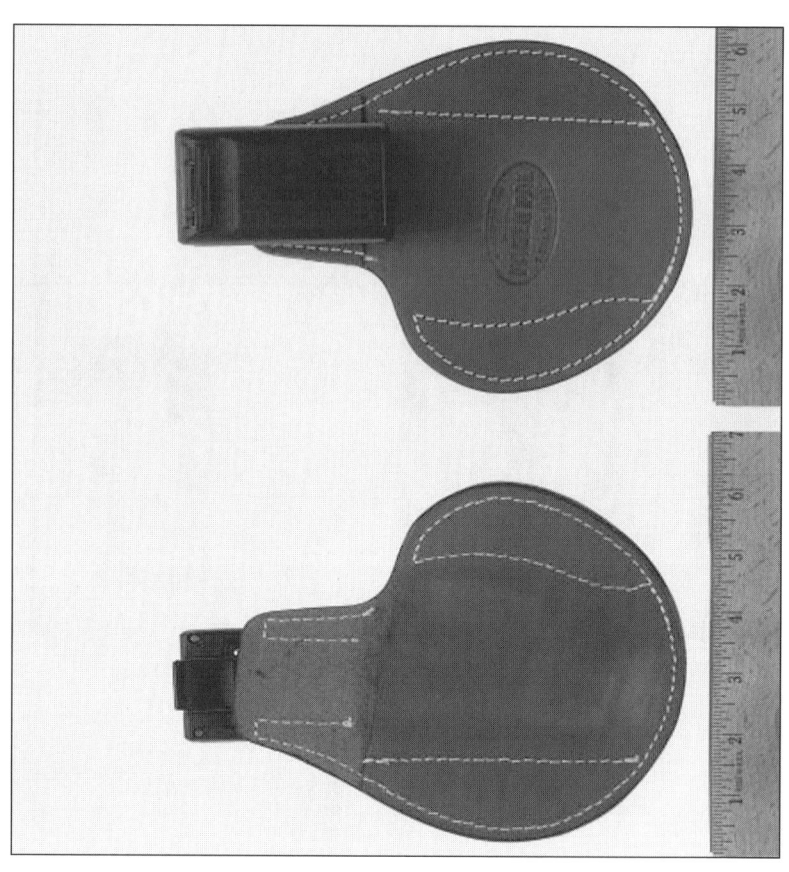

## PAGER PAL HOLSTER

Manufacturer: Pager Pal
Made in USA
Composition: Fake plastic pager with leather holster
Note: This inside-the-pants holster can be used with the provided fake pager or a real one. The holster can also hold small firearms and/or knives.